T0245261

CAMBRIDGE LIBRARY COLLECTION

Books of enduring scholarly value

Technology

The focus of this series is engineering, broadly construed. It covers technological innovation from a range of periods and cultures, but centres on the technological achievements of the industrial era in the West, particularly in the nineteenth century, as understood by their contemporaries. Infrastructure is one major focus, covering the building of railways and canals, bridges and tunnels, land drainage, the laying of submarine cables, and the construction of docks and lighthouses. Other key topics include developments in industrial and manufacturing fields such as mining technology, the production of iron and steel, the use of steam power, and chemical processes such as photography and textile dyes.

Records of a Family of Engineers

Robert Louis Stevenson (1850–94), novelist and poet, was descended from a famous Scottish engineering family. His grandfather Robert, his father Thomas, two uncles and a cousin were all noted engineers, particularly known for their lighthouses. This family history, focusing particularly on his grandfather, was written while R. L. Stevenson was living in Samoa, and was published posthumously in 1912. It first outlines the history of the name 'Stevenson' from the thirteenth century. Chapter 1 begins in the mid-eighteenth century, and records Robert Stevenson's birth in 1772, and his father's death. The young Robert worked with his stepfather for the Northern Lighthouse Board and was its sole Engineer from 1808 to 1843. Chapter 2 describes his experiences in that role. Chapter 3 reproduces substantial extracts from Robert's own diary of the construction (1807–11) of his most famous structure, the Bell Rock lighthouse off Arbroath, which revolutionised lighthouse design.

Cambridge University Press has long been a pioneer in the reissuing of out-of-print titles from its own backlist, producing digital reprints of books that are still sought after by scholars and students but could not be reprinted economically using traditional technology. The Cambridge Library Collection extends this activity to a wider range of books which are still of importance to researchers and professionals, either for the source material they contain, or as landmarks in the history of their academic discipline.

Drawing from the world-renowned collections in the Cambridge University Library, and guided by the advice of experts in each subject area, Cambridge University Press is using state-of-the-art scanning machines in its own Printing House to capture the content of each book selected for inclusion. The files are processed to give a consistently clear, crisp image, and the books finished to the high quality standard for which the Press is recognised around the world. The latest print-on-demand technology ensures that the books will remain available indefinitely, and that orders for single or multiple copies can quickly be supplied.

The Cambridge Library Collection will bring back to life books of enduring scholarly value (including out-of-copyright works originally issued by other publishers) across a wide range of disciplines in the humanities and social sciences and in science and technology.

Records of a Family of Engineers

Robert Louis Stevenson

CAMBRIDGE UNIVERSITY PRESS

Cambridge, New York, Melbourne, Madrid, Cape Town, Singapore,
São Paolo, Delhi, Dubai, Tokyo, Mexico City

Published in the United States of America by Cambridge University Press, New York

www.cambridge.org
Information on this title: www.cambridge.org/9781108026611

This edition first published 1912
This digitally printed version 2011

ISBN 978-1-108-02661-1 Paperback

RECORDS OF
A FAMILY OF ENGINEERS

RECORDS OF A FAMILY OF ENGINEERS

BY

ROBERT LOUIS STEVENSON

LONDON
CHATTO & WINDUS
1912

CONTENTS

RECORDS OF A FAMILY OF ENGINEERS

INTRODUCTION

THE SURNAME OF STEVENSON

From the thirteenth century onwards, the name, under the various disguises of Stevinstoun, Stevensoun, Stevensonne, Stenesone, and Stewinsoune, spread across Scotland from the mouth of the Firth of Forth to the mouth of the Firth of Clyde. Four times at least it occurs as a place-name. There is a parish of Stevenston in Cunningham; a second place of the name in the Barony of Bothwell in Lanark; a third on Lyne, above Drochil Castle; the fourth on the Tyne, near Traprain Law. Stevenson of Stevenson (co. Lanark) swore fealty to Edward I in 1296, and the last of that family died after the Restoration. Stevensons of Hirdmanshiels, in Midlothian, rode in the Bishops' Raid of Aberlady, served as jurors, stood bail for neighbours—Hunter of Polwood, for instance—and became extinct about the same period, or possibly earlier. A Stevenson of Luthrie and another of Pitroddie make their bows, give their names, and vanish. And by the year 1700 it does not appear that any acre of Scots land was vested in any Stevenson.[1]

Here is, so far, a melancholy picture of backward progress, and a family posting towards extinction. But the law (however administered, and I am bound to aver that, in Scotland, 'it couldna weel be waur') acts as a kind of dredge, and with dispassionate impartiality brings up into the light of day, and shows us for a moment, in the jury-box or on the gallows, the creeping things of the past. By these broken glimpses we are able to trace the existence of many other and more inglorious Stevensons, picking a private way through the brawl that makes Scots history. They were members of Parliament for Peebles, Stirling, Pittenweem, Kilrenny, and Inverurie. We find them burgesses of Edinburgh; indwellers in Biggar, Perth, and Dalkeith. Thomas was the forester of Newbattle Park, Gavin was a baker, John a maltman, Francis a chirurgeon, and 'Schir William' a priest. In the feuds of Humes and Heatleys, Cunninghams, Montgomeries, Mures, Ogilvies, and Turnbulls, we find them inconspicuously involved, and apparently getting rather

[1] An error: Stevensons owned at this date the barony of Dolphingston in Haddingtonshire, Montgrennan in Ayrshire, and several other lesser places.

better than they gave. Schir William (reverend gentleman) was cruellie slaughtered on the Links of Kincraig in 1532; James ('in the mill-town of Roberton'), murdered in 1590; Archibald ('in Gallowfarren'), killed with shots of pistols and hagbuts in 1608. Three violent deaths in about seventy years, against which we can only put the case of Thomas, servant to Hume of Cowden Knowes, who was arraigned with his two young masters for the death of the Bastard of Mellerstanes in 1569. John ('in Dalkeith') stood sentry without Holyrood while the banded lords were despatching Rizzio within. William, at the ringing of Perth bell, ran before Gowrie House 'with ane sword, and, entering to the yearde, saw George Craiggingilt with ane twa-handit sword and utheris nychtbouris; at quilk time James Boig cryit ower ane wynds, "Awa hame! ye will all be hangit"'—a piece of advice which William took, and immediately 'depairtit.' John got a maid with child to him in Biggar, and seemingly deserted her; she was hanged on the Castle Hill for infanticide, June 1614; and Martin, elder in Dalkeith, eternally disgraced the name by signing witness in a witch trial, 1661. These are two of our black sheep.[1] Under the Restoration, one Stevenson was a bailie in Edinburgh, and another the lessee of the Canonmills. There were at the same period two

[1] Pitcairn's *Criminal Trials*, at large.—[R. L. S.]

physicians of the name in Edinburgh, one of whom, Dr. Archibald, appears to have been a famous man in his day and generation. The Court had continual need of him; it was he who reported, for instance, on the state of Rumbold; and he was for some time in the enjoyment of a pension of a thousand pounds Scots (about eighty pounds sterling) at a time when five hundred pounds is described as 'an opulent future.' I do not know if I should be glad or sorry that he failed to keep favour; but on 6th January 1682 (rather a cheerless New Year's present) his pension was expunged.[1] There need be no doubt, at least, of my exultation at the fact that he was knighted and recorded arms. Not quite so genteel, but still in public life, Hugh was Under-Clerk to the Privy Council, and liked being so extremely. I gather this from his conduct in September 1681, when, with all the lords and their servants, he took the woful and soul-destroying Test, swearing it 'word by word upon his knees.' And, behold! it was in vain, for Hugh was turned out of his small post in 1684.[2] Sir Archibald and Hugh were both plainly inclined to be trimmers; but there was one witness of the name of Stevenson who held high the banner of the Covenant—John, 'Land-Labourer,[3] in the

[1] Fountainhall's *Decisions*, vol. i. pp. 56, 132, 186, 204, 368.—[R. L. S.]

[2] *Ibid.* pp. 158, 299.—[R. L. S.]

[3] Working farmer: Fr. *laboureur*.

parish of Daily, in Carrick,' that 'eminently pious man.' He seems to have been a poor sickly soul, and shows himself disabled with scrofula, and prostrate and groaning aloud with fever; but the enthusiasm of the martyr burned high within him.

'I was made to take joyfully the spoiling of my goods, and with pleasure for His name's sake wandered in deserts and in mountains, in dens and caves of the earth. I lay four months in the coldest season of the year in a haystack in my father's garden, and a whole February in the open fields not far from Camragen, and this I did without the least prejudice from the night air; one night, when lying in the fields near to the Carrick-Miln, I was all covered with snow in the morning. Many nights have I lain with pleasure in the churchyard of Old Daily, and made a grave my pillow; frequently have I resorted to the old walls about the glen, near to Camragen, and there sweetly rested.' The visible hand of God protected and directed him. Dragoons were turned aside from the bramble-bush where he lay hidden. Miracles were performed for his behoof. 'I got a horse and a woman to carry the child, and came to the same mountain, where I wandered by the mist before; it is commonly known by the name of Kellsrhins: when we came to go up the mountain, there came on a great rain, which we thought was the occasion of the child's weeping, and she wept so bitterly, that all we could do could not divert her from it, so that she was ready to burst. When we got to the top of the mountain, where the Lord had been formerly kind to my soul in prayer, I looked round me for a stone, and espying one, I went and brought it. When the woman with me saw me set down the stone, she smiled, and asked what I was going to do with it. I told her I was going to set it up as my Ebenezer, because hitherto, and in that place, the Lord had formerly helped, and I hoped would yet help. The rain still continuing, the child weeping bitterly, I went to prayer, and no sooner did I cry to God, but the child gave over weeping, and when we got up from prayer, the rain was pouring down on every side, but in the way where we were to go there fell not one drop; the place not rained on was as big as an ordinary avenue.' And so great a saint was the natural butt of Satan's persecutions. 'I retired to the fields for secret prayer about midnight. When I went to pray I was much straitened, and could not get one request, but "Lord pity," "Lord help"; this I came over frequently; at length the terror of Satan fell on me in a high degree, and all I could say even then was—"Lord help." I continued in the duty for some time, notwithstanding of this terror. At length I got up to my feet, and the terror still increased; then the enemy took me by the arm-pits, and seemed to lift me up by my arms. I saw a loch just before me, and I concluded he designed to throw me there by force; and had he got leave to do so, it might have brought a great reproach upon religion.'[1] But it was otherwise ordered, and the cause of piety escaped that danger.[2]

On the whole, the Stevensons may be described as decent, reputable folk, following honest trades—millers, maltsters, and doctors, playing the character parts in the Waverley Novels with propriety, if without distinction; and to an orphan looking about him in the world for a potential ancestry, offering a plain and quite unadorned refuge, equally free from shame and glory. John, the land-labourer, is the one living and memorable figure, and he, alas! cannot possibly be more near than a collateral. It was on August 12, 1678, that he heard Mr. John Welsh on the Craigdowhill, and 'took the heavens, earth, and sun in the firmament that was shining on us, as also the ambassador who made the offer, and *the clerk who raised the psalms,* to witness that I did give myself away to the Lord in a personal and perpetual covenant never to be forgotten'; and already, in 1675, the birth of my direct ascendant was registered in

[1] This John Stevenson was not the only 'witness' of the name; other Stevensons were actually killed during the persecutions, in the Glen of Trool, on Pentland, etc.; and it is very possible that the author's own ancestor was one of the mounted party embodied by Muir of Caldwell, only a day too late for Pentland.

[2] Wodrow Society's *Select Biographies*, vol. ii.—[R. L. S.]

Glasgow. So that I have been pursuing ancestors too far down; and John the land-labourer is debarred me, and I must relinquish from the trophies of my house his *rare soul-strengthening and comforting cordial.* It is the same case with the Edinburgh bailie and the miller of the Canon-mills, worthy man! and with that public character, Hugh the Under-Clerk, and, more than all, with Sir Archibald, the physician, who recorded arms. And I am reduced to a family of inconspicuous maltsters in what was then the clean and handsome little city on the Clyde.

The name has a certain air of being Norse. But the story of Scottish nomenclature is confounded by a continual process of translation and half-translation from the Gaelic which in olden days may have been sometimes reversed. Roy becomes Reid; Gow, Smith. A great Highland clan uses the name of Robertson; a sept in Appin that of Livingstone; Maclean in Glencoe answers to Johnstone at Lockerby. And we find such hybrids as Macalexander for Macallister. There is but one rule to be deduced: that however uncompromisingly Saxon a name may appear, you can never be sure it does not designate a Celt. My great-grandfather wrote the name *Stevenson* but pronounced it *Steenson,* after the fashion of the immortal minstrel in *Redgauntlet*; and this elision of a medial consonant appears a Gaelic process;

and, curiously enough, I have come across no less than two Gaelic forms: *John Macstophane cordinerius in Crossraguel*, 1573, and *William M'Steen* in Dunskeith (co. Ross), 1605. Stevenson, Steenson, Macstophane, M'Steen: which is the original? which the translation? Or were these separate creations of the patronymic, some English, some Gaelic? The curiously compact territory in which we find them seated—Ayr, Lanark, Peebles, Stirling, Perth, Fife, and the Lothians—would seem to forbid the supposition.[1]

'STEVENSON—or according to tradition of one of the proscribed of the clan MacGregor, who was born among the willows or in a hill-side sheep-pen —"Son of my love," a heraldic bar sinister, but history reveals a reason for the birth among the willows far other than the sinister aspect of the name': these are the dark words of Mr. Cosmo Innes; but history or tradition, being interrogated, tells a somewhat tangled tale. The heir of Macgregor of Glenorchy, murdered about 1353 by the Argyll Campbells, appears to have been the original 'Son of my love'; and his more loyal clansmen took the name to fight under. It may be supposed the story of their resistance became popular, and the name in some sort identified with the idea of

[1] Though the districts here named are those in which the name of Stevenson is most common, it is in point of fact far more widespread than the text indicates, and occurs from Dumfries and Berwickshire to Aberdeen and Orkney.

opposition to the Campbells. Twice afterwards, on some renewed aggression, in 1502 and 1552, we find the Macgregors again banding themselves into a sept of 'Sons of my love'; and when the great disaster fell on them in 1603, the whole original legend re-appears, and we have the heir of Alaster of Glenstrae born 'among the willows' of a fugitive mother, and the more loyal clansmen again rallying under the name of Stevenson. A story would not be told so often unless it had some base in fact; nor (if there were no bond at all between the Red Macgregors and the Stevensons) would that extraneous and somewhat uncouth name be so much repeated in the legends of the Children of the Mist.

But I am enabled, by my very lively and obliging correspondent, Mr. George A. Macgregor Stevenson of New York, to give an actual instance. His grandfather, great-grandfather, great-great-grandfather, and great-great-great-grandfather, all used the names of Macgregor and Stevenson as occasion served; being perhaps Macgregor by night and Stevenson by day. The great-great-great-grandfather was a mighty man of his hands, marched with the clan in the 'Forty-five, and returned with *spolia opima* in the shape of a sword, which he had wrested from an officer in the retreat, and which is in the possession of my correspondent to this day. His great-grandson (the grandfather of my correspondent), being converted to Methodism by some wayside preacher, discarded in a moment his name, his old nature, and his political principles, and with the zeal of a proselyte sealed his adherence to the Protestant Succession by baptising his next son George. This George became the publisher and editor of the *Wesleyan Times.* His children were brought up in ignorance of their Highland pedigree; and my correspondent was puzzled to overhear his father speak of him as a true Macgregor, and amazed to find, in rummaging about that peaceful and pious house, the sword of the Hanoverian officer. After he was grown up and was better informed of his descent, 'I frequently asked my father,' he writes, 'why he did not use the name of Macgregor; his replies were significant, and give a picture of the man: "It isn't a good *Methodist* name. *You* can use it, but it will do you no *good.*" Yet the old gentleman, by way of pleasantry, used to announce himself to friends as "Colonel Macgregor."'

Here, then, are certain Macgregors habitually using the name of Stevenson, and at last, under the influence of Methodism, adopting it entirely. Doubtless a proscribed clan could not be particular; they took a name as a man takes an umbrella against a shower; as Rob Roy took Campbell, and his son took Drummond. But this case is different; Stevenson was not taken and left—it was consistently adhered to. It does not in the least follow that all Stevensons are of the clan Alpin; but it does follow that some may be. And I cannot conceal from myself the possibility that James Stevenson in Glasgow, my first authentic ancestor, may have had a Highland *alias* upon his conscience and a claymore in his back parlour.

To one more tradition I may allude, that we are somehow descended from a French barber-surgeon who came to St. Andrews in the service of one of the Cardinal Beatons. No details were added. But the very name of France was so detested in my family for three generations, that I am tempted to suppose there may be something in it.[1]

[1] Mr. J. H. Stevenson is satisfied that these speculations as to a possible Norse, Highland, or French origin are vain. All we know about the engineer family is that it was sprung from a stock of Westland Whigs settled in the latter part of the seventeenth century in the parish of Neilston, as mentioned at the beginning of the next chapter. It may be noted that the Ayrshire parish of Stevenston, the lands of which are said to have received the name in the twelfth century, lies within thirteen miles south-west of this place. The lands of Stevenson in Lanarkshire first mentioned in the next century, in the Ragman Roll, lie within twenty miles east.

CHAPTER I

DOMESTIC ANNALS

It is believed that in 1665, James Stevenson in Nether Carsewell, parish of Neilston, county of Renfrew, and presumably a tenant farmer, married one Jean Keir; and in 1675, without doubt, there was born to these two a son Robert, possibly a maltster in Glasgow. In 1710, Robert married, for a second time, Elizabeth Cumming, and there was born to them, in 1720, another Robert, certainly a maltster in Glasgow. In 1742, Robert the second married Margaret Fulton (Margret, she called herself), by whom he had ten children, among whom were Hugh, born February 1749, and Alan, born June 1752.

With these two brothers my story begins. Their deaths were simultaneous; their lives unusually brief and full. Tradition whispered me in childhood they were the owners of an islet near St. Kitts; and it is certain they had risen to be at the head of considerable interests in the West Indies, which Hugh managed abroad and Alan at home, at an age when others are still curveting a clerk's stool. My kinsman, Mr. Stevenson of Stirling, has heard his father mention that there had been 'something romantic' about Alan's marriage: and, alas! he has forgotten what. It was early at least. His wife was Jean, daughter of David Lillie, a builder in Glasgow, and several times 'Deacon of the Wrights': the date of the marriage has not reached me; but on 8th June 1772, when Robert, the only child of the union, was born, the husband and father had scarce passed, or had not yet attained, his twentieth year. Here was a youth making haste to give hostages to fortune. But this early scene of prosperity in love and business was on the point of closing.

There hung in the house of this young family, and successively in those of my grandfather and father, an oil painting of a ship of many tons burthen. Doubtless the brothers had an interest in the vessel; I was told she had belonged to them outright; and the picture was preserved through years of hardship, and remains to this day in the possession of the family, the only memorial of my great-grandsire Alan. It was on this ship that he sailed on his last adventure, summoned to the West Indies by Hugh. An agent had proved unfaithful on a serious scale; and it used to be told me in my childhood how the brothers pursued him from one island to another in an open boat, were exposed to the pernicious dews of the tropics, and simultaneously struck down. The dates and places of their deaths (now before me) would seem to indicate a more scattered and prolonged pursuit: Hugh, on the 16th April 1774, in Tobago, within sight of Trinidad; Alan, so late as 26th May, and so far away as 'Santt Kittes,' in the Leeward Islands—both, says the family Bible, 'of a fiver' (!). The death of Hugh was probably announced by Alan in a letter, to which we may refer the details of the open boat and the dew Thus, at least, in something like the course of post, both were called away, the one twenty-five, the other twenty-two; their brief generation became extinct, their short-lived house fell with them; and 'in these lawless parts and lawless times'—the words are my grandfather's—their property was stolen or became involved. Many years later, I understand some small recovery to have been made; but at the moment almost the whole means of the family seem to have perished with the young merchants. On the 27th April, eleven days after Hugh Stevenson, twenty-nine before Alan, died David Lillie, the Deacon of the Wrights; so that mother and son were orphaned in one month. Thus, from a few scraps of paper bearing little beyond dates, we construct the outlines of the tragedy that shadowed the cradle of Robert Stevenson.

Jean Lillie was a young woman of strong sense, well fitted to contend with poverty, and of a pious disposition, which it is like that these misfortunes heated. Like so many other widowed Scotswomen, she vowed her son should wag his head in a pulpit; but her means were inadequate to her ambition. A charity school, and some time under a Mr. M'Intyre, 'a famous linguist,' were all she could afford in the way of education to the would-be minister. He learned no Greek; in one place he mentions that the Orations of Cicero were his highest book in Latin; in another that he had 'delighted' in Virgil and Horace; but his delight could never have been scholarly. This appears to have been the whole of his training previous to an event which changed his own destiny and moulded that of his descendants—the second marriage of his mother.

There was a Merchant-Burgess of Edinburgh of the name of Thomas Smith. The Smith pedigree has been traced a little more particularly than the Stevensons', with a similar dearth of illustrious names. One character seems to have appeared, indeed, for a moment at the wings of history: a skipper of Dundee who smuggled over some Jacobite big-wig at the time of the 'Fifteen, and was afterwards drowned in Dundee harbour while going on board his ship. With this exception, the generations of the Smiths present no conceivable interest even to a descendant; and Thomas, of Edinburgh, was the first to issue from respectable obscurity. His father, a skipper out of Broughty Ferry, was drowned at sea while Thomas was still young. He seems to have owned a ship or two—whalers, I suppose, or coasters—and to have been a member of the Dundee Trinity House, whatever that implies. On his death the widow remained in Broughty, and the son came to push his future in Edinburgh. There is a story told of him in the family which I repeat here because I shall have to tell later on a similar, but more perfectly authenticated, experience of his stepson, Robert Stevenson. Word reached Thomas that his mother was unwell, and he prepared to leave for Broughty on the morrow. It was between two and three in the morning, and the early northern daylight was already clear, when he awoke and beheld the curtains at the bed-foot drawn aside and his mother appear in the interval, smile upon him for a moment, and then vanish. The sequel is stereotype; he took the time by his watch, and arrived at Broughty to learn it was the very moment of her death. The incident is at least curious in having happened to such a person—as the tale is being told of him. In all else, he appears as a man, ardent, passionate, practical, designed for affairs and prospering in them far beyond the average. He founded a solid business in lamps and oils, and was the sole proprietor of a concern called the Greenside Company's Works—'a multifarious concern it was,' writes my cousin, Professor Swan, 'of tinsmiths, coppersmiths, brass-founders, blacksmiths, and japanners.' He was also, it seems, a shipowner and underwriter. He built himself 'a land'—Nos. 1 and 2 Baxter's Place, then no such unfashionable neighbourhood—and died, leaving his only son in easy circumstances, and giving to his three surviving daughters portions of five thousand pounds and upwards. There is no standard of success in life; but in one of its meanings, this is to succeed.

In what we know of his opinions, he makes a figure highly characteristic of the time. A high Tory and patriot, a captain—so I find it in my notes—of Edinburgh Spearmen, and on duty in the Castle during the Muir and Palmer troubles, he bequeathed to his descendants a bloodless sword and a somewhat violent tradition, both long preserved. The judge who sat on Muir and Palmer, the famous Braxfield, let fall from the bench the *obiter dictum*—'I never liked the French all my days, but now I hate them.' If Thomas Smith, the Edinburgh Spearman, were in court, he must have been tempted to applaud. The people of that land were his abhorrence; he loathed Buonaparte like Antichrist. Towards the end he fell into a kind of dotage; his family must entertain him with games of tin soldiers, which he took a childish pleasure to array and overset; but those who played with him must be upon their guard, for if his side, which was always that of the English against the French, should chance to be defeated, there would be trouble in Baxter's Place. For these opinions he may almost be said to have suffered. Baptised and brought up in the Church of Scotland, he had, upon some conscientious scruple, joined the communion of the Baptists. Like other Nonconformists, these were inclined to the Liberal side in politics, and, at least in the beginning, regarded Buonaparte as a deliverer. From the time of his joining the Spearmen, Thomas Smith became in consequence a bugbear to his brethren in the faith. 'They that take the sword shall perish with the sword,' they told him; they gave him 'no rest'; 'his position became intolerable'; it was plain he must choose between his political and his religious tenets; and in the last years of his life, about 1812, he returned to the Church of his fathers.

August 1786 was the date of his chief advancement, when, having designed a system of oil lights to take the place of the primitive coal fires before in use, he was dubbed engineer to the newly-formed Board of Northern Lighthouses. Not only were his fortunes bettered by the appointment, but he was introduced to a new and wider field for the exercise of his abilities, and a new way of life highly agreeable to his active constitution. He seems to have rejoiced in the long journeys, and to have combined them with the practice of field sports. 'A tall, stout man coming ashore with his gun over his arm'—so he was described to my father—the only description that has come down to me by a lightkeeper old in the service. Nor did this change come alone. On the 9th July of the same year, Thomas Smith had been left for the second time a widower. As he was still but thirty-three years old, prospering in his affairs, newly advanced in the world, and encumbered at the time with a family of children, five in number, it was natural that he should entertain the notion of another wife. Expeditious in business, he was no less so in his choice; and it was not later than June 1787—for my grandfather is described as still in his fifteenth year—that he married the widow of Alan Stevenson.

The perilous experiment of bringing together two families for once succeeded. Mr. Smith's two eldest daughters, Jean and Janet, fervent in piety, unwearied in kind deeds, were well qualified both to appreciate and to attract the stepmother; and her son, on the other hand, seems to have found immediate favour in the eyes of Mr. Smith. It is, perhaps, easy to exaggerate the ready-made resemblances; the tired woman must have done much to fashion girls who were under ten; the man, lusty and opinionated, must have stamped a strong impression on the boy of fifteen. But the cleavage of the family was too marked, the identity of character and interest produced between the two men on the one hand, and the three women on the other, was too complete to have been the result of influence alone. Particular bonds of union must have pre-existed on each side. And there is no doubt that the man and the boy met with common ambitions, and a common bent, to the practice of that which had not so long before acquired the name of civil engineering.

For the profession which is now so thronged, famous, and influential, was then a thing of yesterday. My grandfather had an anecdote of Smeaton, probably learned from John Clerk of Eldin, their common friend. Smeaton was asked by the Duke of Argyll to visit the West Highland coast for a professional purpose. He refused, appalled, it seems, by the rough travelling. 'You can recommend some other fit person?' asked the Duke. 'No,' said Smeaton, 'I'm sorry I can't.' 'What!' cried the Duke, 'a profession with only one man in it! Pray, who taught you?' 'Why,' said Smeaton, 'I believe I may say I was self-taught, an't please your grace.' Smeaton, at the date of Thomas Smith's third marriage, was yet living; and as the one had grown to the new profession from his place at the instrument-maker's, the other was beginning to enter it by the way of his trade. The engineer of to-day is confronted with a library of acquired results; tables and formulæ to the value of folios full have been calculated and recorded; and the student finds everywhere in front of him the footprints of the pioneers. In the eighteenth century the field was largely unexplored; the engineer must read with his own eyes the face of nature; he arose a volunteer, from the workshop or the mill, to undertake works which were at once inventions and adventures. It was not a science then—it was a living art; and it visibly grew under the eyes and between the hands of its practitioners.

The charm of such an occupation was strongly felt by stepfather and stepson. It chanced that Thomas Smith was a reformer; the superiority of his proposed lamp and reflectors over open fires of coal secured his appointment; and no sooner had he set his hand to the task than the interest of that employment mastered him. The vacant stage on which he was to act, and where all had yet to be created—the greatness of the difficulties, the smallness of the means intrusted him—would rouse a man of his disposition like a call to battle. The lad introduced by marriage under his roof was of a character to sympathise; the public usefulness of the service would appeal to his judgment, the perpetual need for fresh expedients stimulate his ingenuity. And there was another attraction which, in the younger man at least, appealed to, and perhaps first aroused, a profound and enduring sentiment of romance: I mean the attraction of the life. The seas into which his labours carried the new engineer were still scarce charted, the coasts still dark; his way on shore was often far beyond the convenience of any road; the isles in which he must sojourn were still partly savage. He must toss much in boats; he must often adventure on horseback by the dubious bridle-track through unfrequented wildernesses; he must sometimes plant his lighthouse in the very camp of wreckers; and he was continually enforced to the vicissitudes of outdoor life. The joy of my grandfather in this career was strong as the love of woman. It lasted him through youth and manhood, it burned strong in age, and at the approach of death his last yearning was to renew these loved experiences. What he felt himself he continued to attribute to all around him. And to this supposed sentiment in others I find him continually, almost pathetically, appealing; often in vain.

Snared by these interests, the boy seems to have become almost at once the eager confidant and adviser of his new connection; the Church, if he had ever entertained the prospect very warmly, faded from his view; and at the age of nineteen I find him already in a post of some authority, superintending the construction of the lighthouse on the isle of Little Cumbrae, in the Firth of Clyde. The change of aim seems to have caused or been accompanied by a change of character. It sounds absurd to couple the name of my grandfather with the word indolence; but the lad who had been destined from the cradle to the Church, and who had attained the age of fifteen without acquiring more than a moderate knowledge of Latin, was at least no unusual student. And from the day of his charge at Little Cumbrae he steps before us what he remained until the end, a man of the most zealous industry, greedy of occupation, greedy of knowledge, a stern husband of time, a reader, a writer, unflagging in his task of self-improvement. Thenceforward his summers were spent directing works and ruling workmen, now in uninhabited, now in half-savage islands; his winters were set apart, first at the Andersonian Institution, then at the University of Edinburgh to improve himself in mathematics, chemistry, natural history, agriculture, moral philosophy, and logic; a bearded student—although no doubt scrupulously shaved. I find one reference to his years in class which will have a meaning for all who have studied in Scottish Universities. He mentions a recommendation made by the professor of logic. 'The high-school men,' he writes, 'and *bearded men like myself*, were all attention.' If my grandfather were throughout life a thought too studious of the art of getting on, much must be forgiven to the bearded and belated student who looked across, with a sense of difference, at 'the high-school men.' Here was a gulf to be crossed; but already he could feel that he had made a beginning, and that must have been a proud hour when he devoted his earliest earnings to the repayment of the charitable foundation in which he had received the rudiments of knowledge.

In yet another way he followed the example of his father-in-law, and from 1794 to 1807, when the affairs of the Bell Rock made it necessary for him to resign, he served in different corps of volunteers. In the last of these he rose to a position of distinction, no less than captain of the Grenadier Company, and his colonel, in accepting his resignation, entreated he would do them 'the favour of continuing as an honorary member of a corps which has been so much indebted for your zeal and exertions.'

To very pious women the men of the house are apt to appear worldly. The wife, as she puts on her new bonnet before church, is apt to sigh over that assiduity which enabled her husband to pay the milliner's bill. And in the household of the Smiths and Stevensons the women were not only extremely pious, but the men were in reality a trifle worldly. Religious they both were; conscious, like all Scots, of the fragility and unreality of that scene in which we play our uncomprehended parts; like all Scots, realising daily and hourly the sense of another will than ours and a perpetual direction in the affairs of life. But the current of their endeavours flowed in a more obvious channel. They had got on so far; to get on further was their next ambition—to gather wealth, to rise in society, to leave their descendants higher than themselves, to be (in some sense) among the founders of families. Scott was in the same town nourishing similar dreams. But in the eyes of the women these dreams would be foolish and idolatrous.

I have before me some volumes of old letters addressed to Mrs. Smith and the two girls, her favourites, which depict in a strong light their characters and the society in which they moved.

'My very dear and much esteemed Friend,' writes one correspondent, 'this day being the anniversary of our acquaintance, I feel inclined to address you; but where shall I find words to express the fealings of a graitful *Heart,* first to the Lord who graiciously inclined you on this day last year to notice an afflicted Strainger providentially cast in your way far from any Earthly friend? . . . Methinks I shall hear him say unto you, "Inasmuch as ye shewed kindness to my afflicted handmaiden, ye did it unto me."'

This is to Jean; but the same afflicted lady wrote indifferently to Jean, to Janet, and to Mrs. Smith, whom she calls 'my Edinburgh mother.'

It is plain the three were as one person, moving to acts of kindness, like the Graces, inarmed. Too much stress must not be laid on the style of this correspondence; Clarinda survived, not far away, and may have met the ladies on the Calton Hill; and many of the writers appear, underneath the conventions of the period, to be genuinely moved. But what unpleasantly strikes a reader is that these devout unfortunates found a revenue in their devotion. It is everywhere the same tale; on the side of the soft-hearted ladies, substantial acts of help; on the side of the correspondents, affection, italics, texts, ecstasies, and imperfect spelling. When a midwife is recommended, not at all for proficiency in her important art, but because she has 'a sister whom I [the correspondent] esteem and respect, and [who] is a spiritual daughter of my Hon^{d} Father in the Gosple,' the mask seems to be torn off, and the wages of godliness appear too openly. Capacity is a secondary matter in a midwife, temper in a servant, affection in a daughter, and the repetition of a shibboleth fulfils the law. Common decency is at times forgot in the same page with the most sanctified advice and aspiration. Thus I am introduced to a correspondent who appears to have been at the time the housekeeper at Invermay, and who writes to condole with my grandmother in a season of distress. For nearly half a sheet she keeps to the point with an excellent discretion in language; then suddenly breaks out:

'It was fully my intention to have left this at Martinmass, but the Lord fixes the bounds of our habitation. I have had more need of patience in my situation here than in any other, partly from the very violent, unsteady, deceitful temper of the Mistress of the Family, and also from the state of the house. It was in a train of repair when I came here two years ago, and is still in Confusion. There is above six Thousand Pounds' worth of Furniture come from London to be put up when the rooms are completely finished; and then, woe be to the Person who is Housekeeper at Invermay!'

And by the tail of the document, which is torn, I see she goes on to ask the bereaved family to seek her a new place. It is extraordinary that people should have been so deceived in so careless an impostor; that a few sprinkled 'God willings' should have blinded them to the essence of this venomous letter; and that they should have been at the pains to bind it in with others (many of them highly touching) in their memorial of harrowing days. But the good ladies were without guile and without suspicion; they were victims marked for the axe, and the religious impostors snuffed up the wind as they drew near.

I have referred above to my grandmother; it was no slip of the pen: for by an extraordinary arrangement, in which it is hard not to suspect the managing hand of a mother, Jean Smith became the wife of Robert Stevenson. Mrs. Smith had failed in her design to make her son a minister, and she saw him daily more immersed in business and worldly ambition. One thing remained that she might do: she might secure for him a godly wife, that great means of sanctification; and she had two under her hand, trained by herself, her dear friends and daughters both in law and love—Jean and Janet. Jean's complexion was extremely pale, Janet's was florid; my grandmother's nose was straight, my great-aunt's aquiline; but by the sound of the voice, not even a son was able to distinguish one from other. The marriage of a man of twenty-seven and a girl of twenty who have lived for twelve years as brother and sister, is difficult to conceive. It took place, however, and thus in 1799 the family was still further cemented by the union of a representative of the male or worldly element with one of the female and devout.

This essential difference remained unbridged, yet never diminished the strength of their relation. My grandfather pursued his design of advancing in the world with some measure of success; rose to distinction in his calling, grew to be the familiar of members of Parliament, judges of the Court of Session, and 'landed gentlemen'; learned a ready address, had a flow of interesting conversation, and when he was referred to as 'a highly respectable *bourgeois*,' resented the description. My grandmother remained to the end devout and unambitious, occupied with her Bible, her children, and her house; easily shocked, and associating largely with a clique of godly parasites. I do not know if she called in the midwife already referred to; but the principle on which that lady was recommended, she accepted fully. The cook was a godly woman, the butcher a Christian man, and the table suffered. The scene has been often described to me of my grandfather sawing with darkened countenance at some indissoluble joint —'Preserve me, my dear, what kind of a reedy, stringy beast is this?'—of the joint removed, the pudding substituted and uncovered; and of my grandmother's anxious glance and hasty, deprecatory comment, 'Just mismanaged!' Yet with the invincible obstinacy of soft natures, she would adhere to the godly woman and the Christian man, or find others of the same kidney to replace them. One of her confidants had once a narrow escape; an unwieldy old woman, she had fallen from an outside stair in a close of the Old Town; and my grandmother rejoiced to communicate the providential circumstance that a baker had been passing underneath with his bread upon his head. 'I would like to know what kind of providence the baker thought it!' cried my grandfather.

But the sally must have been unique. In all else that I have heard or read of him, so far from criticising, he was doing his utmost to honour and even to emulate his wife's pronounced opinions. In the only letter which has come to my hand of Thomas Smith's, I find him informing his wife that he was 'in time for afternoon church'; similar assurances or cognate excuses abound in the correspondence of Robert Stevenson; and it is comical and pretty to see the two generations paying the same court to a female piety more highly strung: Thomas Smith to the mother of Robert Stevenson—Robert Stevenson to the daughter of Thomas Smith. And if for once my grandfather suffered himself to be hurried, by his sense of humour and justice, into that remark about the case of Providence and the Baker, I should be sorry for any of his children who should have stumbled into the same attitude of criticism. In the apocalyptic style of the housekeeper of Invermay, woe be to that person! But there was no fear; husband and sons all entertained for the pious, tender soul the same chivalrous and moved affection. I have spoken with one who remembered her, and who had been the intimate and equal of her sons, and I found this witness had been struck, as I had been, with a sense of disproportion between the warmth of the adoration felt and the nature of the woman, whether as described or observed. She diligently read and marked her Bible; she was a tender nurse; she had a sense of humour under strong control; she talked and found some amusement at her (or rather at her husband's) dinner-parties. It is conceivable that even my grandmother was amenable to the seductions of dress; at least, I find her husband inquiring anxiously about 'the gowns from Glasgow,' and very careful to describe the toilet of the Princess Charlotte, whom he had seen in church 'in a Pelisse and Bonnet of the same colour of cloth as the Boys' Dress jackets, trimmed with blue satin ribbons; the hat or Bonnet, Mr. Spittal said, was a Parisian slouch, and had a plume of three white feathers.' But all this leaves a blank impression, and it is rather by reading backward in these old musty letters, which have moved me now to laughter and now to impatience, that I glean occasional glimpses of how she seemed to her contemporaries, and trace (at work in her queer world of godly and grateful parasites) a mobile and responsive nature. Fashion moulds us, and particularly women, deeper than we sometimes think; but a little while ago, and, in some circles, women stood or fell by the degree of their appreciation of old pictures; in the early years of the century (and surely with more reason) a character like that of my grandmother warmed, charmed, and subdued, like a strain of music, the hearts of the men of her own household. And there is little doubt that Mrs. Smith, as she looked on at the domestic life of her son and her stepdaughter, and numbered the heads in their increasing nursery, must have breathed fervent thanks to her Creator.

Yet this was to be a family unusually tried; it was not for nothing that one of the godly women saluted Miss Janet Smith as 'a veteran in affliction'; and they were all before middle life experienced in that form of service. By the 1st of January 1808, besides a pair of still-born twins, five children had been born and still survived to the young couple. By the 11th two were gone; by the 28th a third had followed, and the two others were still in danger. In the letters of a former nurserymaid—I give her name, Jean Mitchell, *honoris causa*—we are enabled to feel, even at this distance of time, some of the bitterness of that month of bereavement.

'I have this day received,' she writes to Miss Janet, 'the melancholy news of my dear babys' deaths. My heart is like to break for my dear Mrs. Stevenson. O may she be supported on this trying occasion! I hope her other three babys will be spared to her. O, Miss Smith, did I think when I parted from my sweet babys that I never was to see them more?' 'I received,' she begins her next, 'the mournful news of my dear Jessie's death. I also received the hair of my three sweet babys, which I will preserve as dear to their memorys and as a token of Mr. and Mrs.

Stevenson's friendship and esteem. At my leisure hours, when the children are in bed, they occupy all my thoughts, I dream of them. About two weeks ago I dreamed that my sweet little Jessie came running to me in her usual way, and I took her in my arms. O my dear babys, were mortal eyes permitted to see them in heaven, we would not repine nor grieve for their loss.'

By the 29th of February, the Reverend John Campbell, a man of obvious sense and human value, but hateful to the present biographer, because he wrote so many letters and conveyed so little information, summed up this first period of affliction in a letter to Miss Smith: 'Your dear sister but a little while ago had a full nursery, and the dear blooming creatures sitting around her table filled her breast with hope that one day they should fill active stations in society and become an ornament in the Church below. But ah!'

Near a hundred years ago these little creatures ceased to be, and for not much less a period the tears have been dried. And to this day, looking in these stitched sheaves of letters, we hear the sound of many soft-hearted women sobbing for the lost. Never was such a massacre of the innocents; teething and chincough and scarlet fever and smallpox ran the round; and little Lillies, and Smiths, and Stevensons fell like moths about a candle; and nearly all the sympathetic correspondents deplore and recall the little losses of their own. 'It is impossible to describe the Heavnly looks of the Dear Babe the three last days of his life,' writes Mrs. Laurie to Mrs. Smith. 'Never—never, my dear aunt, could I wish to eface the rememberance of this Dear Child. Never, never, my dear aunt!' And so soon the memory of the dead and the dust of the survivors are buried in one grave.

There was another death in 1812; it passes almost unremarked; a single funeral seemed but a small event to these 'veterans in affliction'; and by 1816 the nursery was full again. Seven little hopefuls enlivened the house; some were growing up; to the elder girl my grandfather already wrote notes in current hand at the tail of his letters to his wife: and to the elder boys he had begun to print, with laborious care, sheets of childish gossip and pedantic applications. Here, for instance, under date of 26th May 1816, is part of a mythological account of London, with a moral for the three gentlemen, Messieurs Alan, Robert, and James Stevenson,' to whom the document is addressed:

'There are many prisons here like Bridewell, for, like other large towns, there are many bad men here as well as many good men. The natives of London are in general not so tall and strong as the people of Edinburgh, because they have not so much pure air, and instead of taking porridge they eat cakes made with sugar and plums. Here you have thousands of carts to draw timber, thousands of coaches to take you to all parts of the town, and thousands of boats to sail on the river Thames. But you must have money to pay, otherwise you can get nothing. Now the way to get money is, become clever men and men of education, by being good scholars.'

From the same absence, he writes to his wife on a Sunday:

'It is now about eight o'clock with me, and I imagine you to be busy with the young folks, hearing the questions [*Anglicé,* catechism], and indulging the boys with a chapter from the large Bible, with their interrogations and your answers in the soundest doctrine. I hope James is getting his verse as usual, and that Mary is not forgetting her little *hymn.* While Jeannie will be reading Wotherspoon, or some other suitable and instructive book, I presume our friend, Aunt Mary, will have just arrived with the news of a *throng kirk* [a crowded church] and a great sermon. You may mention, with my compliments to my mother, that I was at St. Paul's to-day, and attended a very excellent service with Mr. James Lawrie. The text was "Examine and see that ye be in the faith."'

A twinkle of humour lights up this evocation of the distant scene—the humour of happy men and happy homes. Yet it is penned upon the threshold of fresh sorrow. James and Mary—he of the verse and she of the hymn—did not much more than survive to welcome their returning father. On the 25th, one of the godly women writes to Janet:

'My dearest beloved madam, when I last parted from you, you was so affected with your affliction [you? or I?] could think of nothing else. But on Saturday, when I went to inquire after your health, how was I startled to hear that dear James was gone! Ah, what is this? My dear benefactors, doing so much good to many, to the Lord, suddenly to be deprived of their most valued comforts! I was thrown into great perplexity, could do nothing but murmur, why these things were done to such a family. I could not rest, but at midnight, whether spoken [or not] it was presented to my mind—"Those whom ye deplore are walking with me in white." I conclude from this the Lord saying to sweet Mrs. Stevenson: "I gave them to be brought up for me: well done, good and faithful! they are fully prepared, and now I must present them to my father and your father, to my God and your God."'

It would be hard to lay on flattery with a more sure and daring hand. I quote it as a model of a letter of condolence; be sure it would console. Very different, perhaps quite as welcome, is this from a lighthouse inspector to my grandfather:

'In reading your letter the trickling tear ran down my cheeks in silent sorrow for your departed dear ones, my sweet little friends. Well do I remember, and you will call to mind, their little innocent and interesting stories. Often have they come round me and taken me by the hand, but alas! I am no more destined to behold them.'

The child who is taken becomes canonised, and the looks of the homeliest babe seem in the retrospect 'heavenly the three last days of his life.' But it appears that James and Mary had indeed been children more than usually engaging; a record was preserved a long while in the family of their remarks and 'little innocent and interesting stories,' and the blow and the blank were the more sensible.

Early the next month Robert Stevenson must proceed upon his voyage of inspection, part by land, part by sea. He left his wife plunged in low spirits; the thought of his loss, and still more of her concern, was continually present in his mind, and he draws in his letters home an interesting picture of his family relations:

'Windygates Inn, Monday (Postmark July 16th).

'My dearest Jeannie,—While the people of the inn are getting me a little bit of something to eat, I sit down to tell you that I had a most excellent passage across the water, and got to Wemyss at mid-day. I hope the children will be very good, and that Robert will take a course with you to learn his Latin lessons daily; he may, however, read English in company. Let them have strawberries on Saturdays.'

'Westhaven, 17th July.

'I have been occupied to-day at the harbour of Newport, opposite Dundee, and am this far on my way to Arbroath. You may tell the boys that I slept last night in Mr. Steadman's tent. I found my bed rather hard, but the lodgings were otherwise extremely comfortable. The encampment is on the Fife side of the Tay, immediately opposite to Dundee. From the door of the tent you command the most beautiful view of the Firth, both up and down, to a great extent. At night all was serene and still, the sky presented the most beautiful appearance of bright stars, and the morning was ushered in with the song of many little birds.'

'Aberdeen, July 19th.

'I hope, my dear, that you are going out of doors regularly and taking much exercise. I would have you to *make the markets daily*—and by all means to take a seat in the coach once or twice in the week and see what is going on in town [The family were at the sea-side.] It will be good not to be too great a stranger to the house. It will be rather painful at first, but as it is to be done, I would have you not to be too strange to the house in town.

'Tell the boys that I fell in with a soldier—his name is Henderson—who was twelve years with Lord Wellington and other commanders. He returned very lately with only eightpence-halfpenny in his pocket, and found his father and mother both in life, though they had never heard from him, nor he from them. He carried my great-coat and umbrella a few miles.'

'Fraserburgh, July 20th.

'Fraserburgh is the same dull place which [Auntie] Mary and Jeannie found it. As I am travelling along the coast which they are acquainted with, you had better cause Robert bring down the map from Edinburgh; and it will be a good exercise in geography for the young folks to trace my course. I hope they have entered upon the writing. The library will afford abundance of excellent books, which I wish you would employ a little. I hope you are doing me the favour to go much out with the boys, which will do you much good and prevent them from getting so very much overheated.'

[*To the Boys—Printed.*]

When I had last the pleasure of writing to you, your dear little brother James and your sweet little sister Mary were still with us. But it has pleased God to remove them to another and a better world, and we must submit to the will of Providence. I must, however, request of you to think sometimes upon them, and to be very careful not to do anything that will displease or vex your mother. It is therefore proper that you do not roamp [Scottish indeed] too much about, and that you learn your lessons.

'I went to Fraserburgh and visited Kinnaird Head Lighthouse, which I found in good order. All this time I travelled upon good roads, and paid many a toll-man by the way; but from Fraserburgh to Banff there is no toll-bars, and the road is so bad that I had to walk up and down many a hill, and for want of bridges the horses had to drag the chaise up to the middle of the wheels in water. At Banff I saw a large ship of 300 tons lying on the sands upon her beam-ends, and a wreck for want of a good harbour. Captain Wilson—to whom I beg my compliments—will show you a ship of 300 tons. At the towns of Macduff, Banff, and Portsoy, many of the houses are built of marble, and the rocks on this part of the coast or sea-side are marble. But, my dear Boys, unless marble be polished and dressed, it is a very coarse-looking stone, and has no more beauty than common rock. As a proof of this, ask the favour of your mother to take you to Thomson's Marble Works in South Leith, and you will see marble in all its stages, and perhaps you may there find Portsoy marble! The use I wish to make of this is to tell you that, without education, a man is just like a block of rough, unpolished marble. Notice, in proof of this, how much Mr. Neill and Mr. M'Gregor [the tutor] know, and observe how little a man knows who is not a good scholar. On my way to Fochabers I passed through many thousand acres of Fir timber, and saw many deer running in these woods.'

[*To Mrs. Stevenson.*]

'*Inverness, July 21st.*

'I propose going to church in the afternoon, and as I have breakfasted late, I shall afterwards take a walk, and dine about six o'clock. I do not know who is the clergyman here, but I shall think of you all. I travelled in the mail-coach [from Banff] almost alone. While it was daylight I kept the top, and the passing along a country I had never before seen was a considerable amusement. But, my dear, you are all much in my thoughts, and many are the objects which recall the recollection of our tender and engaging children we have so recently lost. We must not, however, repine. I could not for a moment wish any change of circumstances in their case; and in every comparative view of their state, I see the Lord's goodness in removing them from an evil world to an abode of bliss; and I must earnestly hope that you may be enabled to take such a view of this affliction as to live in the happy prospect of our all meeting again to part no more—and that under such considerations you are getting up your spirits. I wish you would walk about, and by all means go to town, and do not sit much at home.'

'*Inverness, July 23rd.*

'I am duly favoured with your much-valued letter, and I am happy to find that you are so much with my mother, because that sort of variety has a tendency to occupy the mind, and to keep it from brooding too much upon one subject. Sensibility and tenderness are certainly two of the most interesting and pleasing qualities of the mind. These qualities are also none of the least of the many endearingments of the female character. But if that kind of sympathy and pleasing melancholy, which is familiar to us under distress, be much indulged, it becomes habitual, and takes such a hold of the mind as to absorb all the other affections, and unfit us for the duties and proper enjoyments of life. Resignation sinks into a kind of peevish discontent. I am far, however, from thinking there is the least danger of this in your case, my dear; for you have been on all occasions enabled to look upon the fortunes of this life as under the direction of a higher power, and have always preserved that propriety and consistency of conduct in all circumstances which endears your example to your family in particular, and to your friends. I am therefore, my dear, for you to go out much, and to go to the house up-stairs [he means to go up-stairs in the house, to visit the place of the dead children], and to put yourself in the way of the visits of your friends. I wish you would call on the Miss Grays, and it would be a good thing upon a Saturday to dine with my mother, and take Meggy and all the family with you, and let them have their strawberries in town. The tickets of one of the *old-fashioned coaches* would take you all up, and if the evening were good, they could all walk down, excepting Meggy and little David.'

'Inverness, July 25th, 11 p.m.

'Captain Wemyss, of Wemyss, has come to Inverness to go the voyage with me, and as we are sleeping in a double-bedded room, I must no longer transgress. You must remember me the best way you can to the children.'

'On board of the Lighthouse Yacht, July 29th.

'I got to Cromarty yesterday about mid-day, and went to church. It happened to be the sacrament there, and I heard a Mr. Smith at that place conclude the service with a very suitable exhortation. There seemed a great concourse of people, but they had rather an unfortunate day for them at the tent, as it rained a good deal. After drinking tea at the inn, Captain Wemyss accompanied me on board, and we sailed about eight last night. The wind at present being rather a beating one, I think I shall have an opportunity of standing into the bay of Wick, and leaving this letter to let you know my progress and that I am well.'

'Lighthouse Yacht, Stornoway, August 4th.

'To-day we had prayers on deck as usual when at sea. I read the 14th chapter, I think, of Job. Captain Wemyss has been in the habit of doing this on board his own ship, agreeably to the Articles of War. Our passage round the Cape [Cape Wrath] was rather a cross one, and as the wind was northerly, we had a pretty heavy sea, but upon the whole have made a good passage, leaving many vessels behind us in Orkney. I am quite well, my dear; and Captain Wemyss, who has much spirit, and who is much given to observation, and a perfect enthusiast in his profession, enlivens the voyage greatly. Let me entreat you to move about much, and take a walk with the boys to Leith. I think they have still many places to see there, and I wish you would indulge them in this respect. Mr. Scales is the best person I know for showing them the sailcloth-weaving, etc., and he would have great pleasure in undertaking this. My dear, I trust soon to be with you, and that through the goodness of God we shall meet all well.

'There are two vessels lying here with emigrants for America, each with eighty people on board, at all ages, from a few days to upwards of sixty! Their prospects must be very forlorn to go with a slender purse for distant and unknown countries.'

'*Lighthouse Yacht, off Greenock, Aug.* 18*th.*

'It was after *church-time* before we got here, but we had prayers upon deck on the way up the Clyde. This has, upon the whole, been a very good voyage, and Captain Wemyss, who enjoys it much, has been an excellent companion; we met with pleasure, and shall part with regret.'

Strange that, after his long experience, my grandfather should have learned so little of the attitude and even the dialect of the spiritually-minded; that after forty-four years in a most religious circle, he could drop without sense of incongruity from a period of accepted phrases to 'trust his wife was *getting up her spirits*,' or think to reassure her as to the character of Captain Wemyss by mentioning that he had read prayers on the deck of his frigate '*agreeably to the Articles of War*'! Yet there is no doubt—and it is one of the most agreeable features of the kindly series—that he was doing his best to please, and there is little doubt that he succeeded. Almost all my grandfather's private letters have been destroyed. This correspondence has not only been preserved entire, but stitched up in the same covers with the works of the godly women, the Reverend John Campbell, and the painful Mrs. Ogle. I did not think to mention the good dame, but she comes in usefully as an example. Amongst the treasures of the ladies of my family, her letters have been honoured with a volume to themselves. I read about a half of them myself; then handed over the task to one of stauncher resolution, with orders to communicate any fact that should be found to illuminate these pages. Not one was found; it was her only art to communicate by post second-rate sermons at second-hand; and such, I take it, was the correspondence in which my grandmother delighted. If I am right, that of Robert Stevenson, with his quaint smack of the contemporary 'Sandford and Merton,' his interest in the whole page of experience, his perpetual quest, and fine scent of all that seems romantic to a boy, his needless pomp of language, his excellent good sense, his unfeigned, unstained, unwearied human kindliness, would seem to her, in a comparison, dry and trivial and worldly. And if these letters were by an exception cherished and preserved, it would be for one or both of two reasons—because they dealt with and were bitter-sweet reminders of a time of sorrow; or because she was pleased, perhaps touched, by the writer's guileless efforts to seem spiritually-minded.

After this date there were two more births and two more deaths, so that the number of the family remained unchanged; in all five children survived to reach maturity and to outlive their parents.

CHAPTER II

THE SERVICE OF THE NORTHERN LIGHTS

I

It were hard to imagine a contrast more sharply defined than that between the lives of the men and women of this family: the one so chambered, so centred in the affections and the sensibilities; the other so active, healthy, and expeditious. From May to November, Thomas Smith and Robert Stevenson were on the mail, in the saddle, or at sea; and my grandfather, in particular, seems to have been possessed with a demon of activity in travel. In 1802, by direction of the Northern Lighthouse Board, he had visited the coast of England from St. Bees, in Cumberland, and round by the Scilly Islands to some place undecipherable by me; in all a distance of 2500 miles. In 1806 I find him starting 'on a tour round the south coast of England, from the Humber to the Severn.' Peace was not long declared ere he found means to visit Holland, where he was in time to see, in the navy-yard at Helvoetsluys, 'about twenty of Bonaparte's *English flotilla* lying in a state of decay, the object of curiosity to Englishmen.' By 1834 he seems to have been acquainted with the coast of France from Dieppe to Bordeaux; and a main part of his duty as Engineer to the Board of Northern Lights was one round of dangerous and laborious travel.

In 1786, when Thomas Smith first received the appointment, the extended and formidable coast of Scotland was lighted at a single point—the Isle of May, in the jaws of the Firth of Forth, where, on a tower already a hundred and fifty years old, an open coal-fire blazed in an iron chauffer. The whole archipelago, thus nightly plunged in darkness, was shunned by sea-going vessels, and the favourite courses were north about Shetland and west about St. Kilda. When the Board met, four new lights formed the extent of their intentions —Kinnaird Head, in Aberdeenshire, at the eastern elbow of the coast; North Ronaldsay, in Orkney, to keep the north and guide ships passing to the south'ard of Shetland; Island Glass, on Harris, to mark the inner shore of the Hebrides and illuminate the navigation of the Minch; and the Mull of Kintyre. These works were to be attempted against obstacles, material and financial, that might have staggered the most bold. Smith had no ship at his command till 1791; the roads in those outlandish quarters where his business lay were scarce passable when they existed, and the tower on the Mull of Kintyre stood eleven months unlighted while the apparatus toiled and foundered by the way among rocks and mosses. Not only had towers to be built and apparatus transplanted; the supply of oil must be maintained, and the men fed, in the same inaccessible and distant scenes; a whole service, with its routine and hierarchy, had to be called out of nothing; and a new trade (that of lightkeeper) to be taught, recruited, and organised. The funds of the Board were at the first laughably inadequate. They embarked on their career on a loan of twelve hundred pounds, and their income in 1789, after relief by a fresh Act of Parliament, amounted to less than three hundred. It must be supposed that the thoughts of Thomas Smith, in these early years, were sometimes coloured with despair; and since he built and lighted one tower after another, and created and bequeathed to his successors the elements of an excellent administration, it may be conceded that he was not after all an unfortunate choice for a first engineer.

War added fresh complications. In 1794 Smith came ‘ very near to be taken ’ by a French squadron. In 1813 Robert Stevenson was cruising about the neighbourhood of Cape Wrath in the immediate fear of Commodore Rogers. The men, and especially the sailors, of the lighthouse service must be protected by a medal and ticket from the brutal activity of the press-gang. And the zeal of volunteer patriots was at times embarrassing.

'I set off on foot,' writes my grandfather, 'for Marazion, a town at the head of Mount's Bay, where I was in hopes of getting a boat to freight. I had just got that length, and was making the necessary inquiry, when a young man, accompanied by several idle-looking fellows, came up to me, and in a hasty tone said, "Sir, in the king's name I seize your person and papers." To which I replied that I should be glad to see his authority, and know the reason of an address so abrupt. He told me the want of time prevented his taking regular steps, but that it would be necessary for me to return to Penzance, as I was suspected of being a French spy. I proposed to submit my papers to the nearest Justice of Peace, who was immediately applied to, and came to the inn where I was. He seemed to be greatly agitated, and quite at a loss how to proceed. The complaint preferred against me was "that I had examined the Longships Lighthouse with the most minute attention, and was no less particular in my inquiries at the keepers of the lighthouse regarding the sunk rocks lying off the Land's End, with the sets of the currents and tides along the coast: that I seemed particularly to regret the situation of the rocks called the Seven Stones, and the loss of a beacon which the Trinity Board had caused to be fixed on the Wolf Rock; that I had taken notes of the bearings of several sunk rocks, and a drawing of the lighthouse, and of Cape Cornwall. Further, that I had refused the honour of Lord Edgecombe's invitation to dinner, offering as an apology that I had some particular business on hand."'

My grandfather produced in answer his credentials and letter of credit; but the justice, after perusing them, very gravely observed that they were "musty bits of paper,"' and proposed to maintain the arrest. Some more enlightened magistrates at Penzance relieved him of suspicion and left him at liberty to pursue his journey,—'which I did with so much eagerness,' he adds, 'that I gave the two coal lights on the Lizard only a very transient look.'

Lighthouse operations in Scotland differed essentially in character from those in England. The English coast is in comparison a habitable, homely place, well supplied with towns; the Scottish presents hundreds of miles of savage islands and desolate moors. The Parliamentary committee of 1834, profoundly ignorant of this distinction, insisted with my grandfather that the work at the various stations should be let out on contract 'in the neighbourhood,' where sheep and deer, and gulls and cormorants, and a few ragged gillies, perhaps crouching in a bee-hive house, made up the only neighbours. In such situations repairs and improvements could only be overtaken by collecting (as my grandfather expressed it) a few 'lads,' placing them under charge of a foreman, and despatching them about the coast as occasion served. The particular danger of these seas increased the difficulty. The course of the lighthouse tender lies amid iron-bound coasts, among tide-races, the whirlpools of the Pentland Firth, flocks of islands, flocks of reefs, many of them uncharted. The aid of steam was not yet. At first in random coasting sloop, and afterwards in the cutter belonging to the service, the engineer must ply and run amongst these multiplied dangers, and sometimes late into the stormy autumn. For pages together my grandfather's diary preserves a record of these rude experiences; of hard winds and rough seas; and of 'the try-sail and storm-jib, those old friends which I never like to see.' They do not tempt to quotation, but it was the man's element, in which he lived, and delighted to live, and some specimen must be presented. On Friday, September 10th, 1830, the *Regent* lying in Lerwick Bay, we have this entry: 'The gale increases, with continued rain.' On the morrow, Saturday, 11th, the weather appeared to moderate, and they put to sea, only to be driven by evening into Levenswick. There they lay, 'rolling much,' with both anchors ahead and the square yard on deck, till the morning of Saturday, 18th. Saturday and Sunday they were plying to the southward with a 'strong breeze and a heavy sea,' and on Sunday evening anchored in Otterswick. 'Monday, 20th, it blows so fresh that we have no communication with the shore. We see Mr. Rome on the beach, but we cannot communicate with him. It blows "mere fire," as the sailors express it.' And for three days more the diary goes on with tales of davits unshipped, high seas, strong gales from the southward, and the ship driven to refuge in Kirkwall or Deer Sound. I have many a passage before me to transcribe, in which my grandfather draws himself as a man of minute and anxious exactitude about details. It must not be forgotten that these voyages in the tender were the particular pleasure and reward of his existence; that he had in him a reserve of romance which carried him delightedly over these hardships and perils; that to him it was 'great gain' to be eight nights and seven days in the savage bay of Levenswick—to read a book in the much agitated cabin—to go on deck and hear the gale scream in his ears, and see the landscape dark with rain, and the ship plunge at her two anchors—and to turn in at night and wake again at morning, in his narrow berth, to the clamorous and continued voices of the gale.

His perils and escapes were beyond counting. I shall only refer to two: the first, because of the impression made upon himself; the second, from the incidental picture it presents of the north islanders. On the 9th October 1794 he took passage from Orkney in the sloop *Elizabeth* of Stromness. She made a fair passage till within view of Kinnaird Head, where, as she was becalmed some three miles in the offing, and wind seemed to threaten from the south-east, the captain landed him, to continue his journey more expeditiously ashore. A gale immediately followed, and the *Elizabeth* was driven back to Orkney and lost with all hands. The second escape I have been in the habit of hearing related by an eye-witness, my own father, from the earliest days of childhood. On a September night, the *Regent* lay in the Pentland Firth in a fog and a violent and windless swell. It was still dark, when they were alarmed by the sound of breakers, and an anchor was immediately let go. The peep of dawn discovered them swinging in desperate proximity to the Isle of Swona [1] and the surf bursting close under their stern. There was in this place a hamlet of the inhabitants, fisher-folk and wreckers; their huts stood close about the head of the beach. All slept; the doors were closed, and there was no smoke, and the anxious watchers on board ship seemed to contemplate a village of the dead. It was thought possible to launch a boat and tow the *Regent* from her place of danger; and with this view a signal of distress was made and a gun fired with a red-hot poker from the galley. Its detonation awoke the sleepers. Door after door was opened, and in the grey light of the morning fisher after fisher was

[1] This is only a probable hypothesis; I have tried to identify my father's anecdote in my grandfather's diary, and may very well have been deceived.—[R. L. S.]

seen to come forth, yawning and stretching himself, nightcap on head. Fisher after fisher, I wrote, and my pen tripped; for it should rather stand wrecker after wrecker. There was no emotion, no animation, it scarce seemed any interest; not a hand was raised; but all callously awaited the harvest of the sea, and their children stood by their side and waited also. To the end of his life, my father remembered that amphitheatre of placid spectators on the beach; and with a special and natural animosity, the boys of his own age. But presently a light air sprang up, and filled the sails, and fainted, and filled them again; and little by little the *Regent* fetched way against the swell, and clawed off shore into the turbulent firth.

The purpose of these voyages was to effect a landing on open beaches or among shelving rocks, not for persons only, but for coals and food, and the fragile furniture of light-rooms. It was often impossible. In 1831 I find my grandfather 'hovering for a week' about the Pentland Skerries for a chance to land; and it was almost always difficult. Much knack and enterprise were early developed among the seamen of the service; their management of boats is to this day a matter of admiration; and I find my grandfather in his diary depicting the nature of their excellence in one happily descriptive phrase, when he remarks that Captain Soutar had landed 'the small stores and nine casks of oil *with all the activity of a smuggler.*' And it was one thing to land, another to get on board again. I have here a passage from the diary, where it seems to have been touch-and-go. 'I landed at Tarbetness, on the eastern side of the point, in *a mere gale or blast of wind* from west-south-west, at 2 p.m. It blew so fresh that the captain, in a kind of despair, went off to the ship, leaving myself and the steward ashore. While I was in the light-room, I felt it shaking and waving, not with the tremor of the Bell Rock, but with the *waving of a tree*! This the light-keepers seemed to be quite familiar to, the principal keeper remarking that "it was very pleasant," perhaps meaning interesting or curious. The captain worked the vessel into smooth water with admirable dexterity and I got on board again about 6 p.m. from the other side of the point.' But not even the dexterity of Soutar could prevail always; and my grandfather must at times have been left in strange berths and with but rude provision. I may instance the case of my father, who was storm-bound three days upon an islet, sleeping in the uncemented and unchimneyed houses of the islanders, and subsisting on a diet of nettle-soup and lobsters.

The name of Soutar has twice escaped my pen, and I feel I owe him a vignette. Soutar first attracted notice as mate of a praam at the Bell

Rock, and rose gradually to be captain of the *Regent*. He was active, admirably skilled in his trade, and a man incapable of fear. Once, in London, he fell among a gang of confidence-men, naturally deceived by his rusticity and his prodigious accent. They plied him with drink—a hopeless enterprise, for Soutar could not be made drunk; they proposed cards, and Soutar would not play. At last, one of them, regarding him with a formidable countenance, inquired if he were not frightened? 'I 'm no' very easy fleyed,' replied the captain. And the rooks withdrew after some easier pigeon. So many perils shared, and the partial familiarity of so many voyages, had given this man a stronghold in my grandfather's estimation; and there is no doubt but he had the art to court and please him with much hypocritical skill. He usually dined on Sundays in the cabin. He used to come down daily after dinner for a glass of port or whisky, often in his full rig of sou'-wester, oilskins, and long boots; and I have often heard it described how insinuatingly he carried himself on these appearances, artfully combining the extreme of deference with a blunt and seamanlike demeanour. My father and uncles, with the devilish penetration of the boy, were far from being deceived; and my father, indeed, was favoured with an object-lesson not to be mistaken. He had crept one rainy night into an apple-barrel on deck, and from this place of ambush overheard Soutar and a comrade conversing in their oilskins. The smooth sycophant of the cabin had wholly disappeared, and the boy listened with wonder to a vulgar and truculent ruffian. Of Soutar, I may say *tantum vidi,* having met him in the Leith docks now more than thirty years ago, when he abounded in the praises of my grandfather, encouraged me (in the most admirable manner) to pursue his footprints, and left impressed for ever on my memory the image of his own Bardolphian nose. He died not long after.

The engineer was not only exposed to the hazards of the sea; he must often ford his way by land to remote and scarce accessible places, beyond reach of the mail or the post-chaise, beyond even the tracery of the bridle-path, and guided by natives across bog and heather. Up to 1807 my grandfather seems to have travelled much on horseback; but he then gave up the idea—'such,' he writes with characteristic emphasis and capital letters, 'is the Plague of Baiting.' He was a good pedestrian; at the age of fifty-eight I find him covering seventeen miles over the moors of the Mackay country in less than seven hours, and that is not bad travelling for a scramble. The piece of country traversed was already a familiar track, being that between Loch Eriboll and Cape Wrath; and I think I can scarce do better than reproduce from the diary some traits of his first visit. The tender lay in Loch Eriboll; by five in the morning they sat down to breakfast on board; by six they were ashore—my grandfather, Mr. Slight an assistant, and Soutar of the jolly nose, and had been taken in charge by two young gentlemen of the neighbourhood and a pair of gillies. About noon they reached the Kyle of Durness and passed the ferry. By half-past three they were at Cape Wrath—not yet known by the emphatic abbreviation of 'The Cape'—and beheld upon all sides of them unfrequented shores, an expanse of desert moor, and the high-piled Western Ocean. The site of the tower was chosen. Perhaps it is by inheritance of blood, but I know few things more inspiriting than this location of a lighthouse in a designated space of heather and air, through which the sea-birds are still flying. By 9 p.m. the return journey had brought them again to the shores of the Kyle. The night was dirty, and as the sea was high and the ferry-boat small, Soutar and Mr. Stevenson were left on the far side, while the rest of the party embarked and were received into the darkness. They made, in fact, a safe though an alarming passage; but the ferryman refused to repeat the adventure; and my grandfather and the captain long paced the beach, impatient for their turn to pass, and tormented with rising anxiety as to the fate of their companions. At length they sought the shelter of a shepherd's house. 'We had miserable up-putting,' the diary continues, 'and on both sides of the ferry much anxiety of mind. Our beds were clean straw, and but for the circumstance of the boat, I should have slept as soundly as ever I did after a walk through moss and mire of sixteen hours.'

To go round the lights, even to-day, is to visit past centuries. The tide of tourists that flows yearly in Scotland, vulgarising all where it approaches, is still defined by certain barriers. It will be long ere there is a hotel at Sumburgh or a hydropathic at Cape Wrath; it will be long ere any *char-à-banc*, laden with tourists, shall drive up to Barra Head or Monach, the Island of the Monks. They are farther from London than St. Petersburg, and except for the towers, sounding and shining all night with fog-bells and the radiance of the light-room, glittering by day with the trivial brightness of white paint, these island and moorland stations seem inaccessible to the civilisation of to-day, and even to the end of my grandfather's career the isolation was far greater. There ran no post at all in the Long Island; from the lighthouse on Barra Head a boat must be sent for letters as far as Tobermory, between sixty and seventy miles of open sea; and the posts of Shetland, which had surprised Sir Walter Scott in 1814, were still unimproved in 1833, when my grandfather reported on the subject. The group contained at the time a population of 30,000 souls, and enjoyed a trade which had increased in twenty years sevenfold, to between three and four thousand tons. Yet the mails were despatched and received by chance coasting vessels at the rate of a penny a letter; six and eight weeks often elapsed between opportunities, and when a mail was to be made up, sometimes at a moment's notice, the bellman was sent hastily through the streets of Lerwick. Between Shetland and Orkney, only seventy miles apart, there was 'no trade communication whatever.'

Such was the state of affairs, only sixty years ago, with the three largest clusters of the Scottish Archipelago; and forty-seven years earlier, when Thomas Smith began his rounds, or forty-two, when Robert Stevenson became conjoined with him in these excursions, the barbarism was deep, the people sunk in superstition, the circumstances of their life perhaps unique in history. Lerwick and Kirkwall, like Guam or the Bay of Islands, were but barbarous ports where whalers called to take up and to return experienced seamen. On the outlying islands the clergy lived isolated, thinking other thoughts, dwelling in a different country from their parishioners, like missionaries in the South Seas. My grandfather's unrivalled treasury of anecdote was never written down;

it embellished his talk while he yet was, and died with him when he died; and such as have been preserved relate principally to the islands of Ronaldsay and Sanday, two of the Orkney group. These bordered on one of the water-highways of civilisation; a great fleet passed annually in their view, and of the shipwrecks of the world they were the scene and cause of a proportion wholly incommensurable to their size. In one year, 1798, my grandfather found the remains of no fewer than five vessels on the isle of Sanday, which is scarcely twelve miles long.

'Hardly a year passed,' he writes, 'without instances of this kind; for, owing to the projecting points of this strangely formed island, the lowness and whiteness of its eastern shores, and the wonderful manner in which the scanty patches of land are intersected with lakes and pools of water, it becomes, even in daylight, a deception, and has often been fatally mistaken for an open sea. It had even become proverbial with some of the inhabitants to observe that "if wrecks were to happen, they might as well be sent to the poor isle of Sanday as anywhere else." On this and the neighbouring islands the inhabitants had certainly had their share of wrecked goods, for the eye is presented with these melancholy remains in almost every form. For example, although quarries are to be met with generally in these islands, and the stones are very suitable for building dykes (*Anglicé*, walls), yet instances occur of the land being enclosed, even to a considerable extent, with ship-timbers. The author has actually seen a park (*Anglicé*, meadow) paled round chiefly with cedar-wood and mahogany from the wreck of a Honduras-built ship; and in one island, after the wreck of a ship laden with wine, the inhabitants have been known to take claret to their barley-meal porridge. On complaining to one of the pilots of the badness of his boat's sails, he replied to the author with some degree of pleasantry, "Had it been His will that you came na' here wi' your lights, we might 'a' had better sails to our boats, and more o' other things." It may further be mentioned that when some of Lord Dundas's farms are to be let in these islands a competition takes place for the lease, and it is *bona fide* understood that a much higher rent is paid than the lands would otherwise give were it not for the chance of making considerably by the agency and advantages attending shipwrecks on the shores of the respective farms.'

The people of North Ronaldsay still spoke Norse, or, rather, mixed it with their English. The walls of their huts were built to a great thickness of rounded stones from the sea-beach; the roof flagged, loaded with earth, and perforated by a single hole for the escape of smoke. The grass grew beautifully green on the flat house-top, where the family would assemble with their dogs and cats, as on a pastoral lawn; there were no windows, and in my grandfather's expression, 'there was really no demonstration of a house unless it were the diminutive door.' He once landed on Ronaldsay with two friends. 'The inhabitants crowded and pressed so much upon the strangers that the bailiff, or resident factor of the island, blew with his ox-horn, calling out to the natives to stand off and let the gentlemen come forward to the laird; upon which one of the islanders, as spokesman, called out, "God ha'e us, man! thou needsna mak' sic a noise. It 's no' every day we ha'e *three hatted men* on our isle."' When the Surveyor of Taxes came (for the first time, perhaps) to Sanday, and began in the King's name to complain of the unconscionable swarms of dogs, and to menace the inhabitants with taxation, it chanced that my grandfather and his friend, Dr. Patrick Neill, were received by an old lady in a Ronaldsay hut. Her hut, which was similar to the model described, stood on a Ness, or point of land jutting into the sea. They were made welcome in the firelit cellar, placed 'in *casey* or straw-worked chairs, after the Norwegian fashion, with arms, and a canopy overhead,' and given milk in a wooden dish. These hospitalities attended to, the old lady turned at once to Dr. Neill, whom she took for the Surveyor of Taxes. 'Sir,' said she, 'gin ye 'll tell the King that I canna keep the Ness free o' the Bangers (sheep) without twa hun's, and twa guid hun's too, he 'll pass me threa the tax on dugs.'

This familiar confidence, these traits of engaging simplicity, are characters of a secluded people. Mankind—and, above all, islanders—come very swiftly to a bearing, and find very readily, upon one convention or another, a tolerable corporate life. The danger is to those from without, who have not grown up from childhood in the islands, but appear suddenly in that narrow horizon, life-sized apparitions. For these no bond of humanity exists, no feeling of kinship is awakened by their peril; they will assist at a shipwreck, like the fisher-folk of Lunga, as spectators, and when the fatal scene is over, and the beach strewn with dead bodies, they will fence their fields with mahogany, and, after a decent grace, sup claret to their porridge. It is not wickedness: it is scarce evil; it is only, in its highest power, the sense of isolation and the wise disinterestedness of feeble and poor races. Think how many viking ships had sailed by these islands in the past, how many vikings had landed, and raised turmoil, and broken up the barrows of the dead, and carried off the wines of the living; and blame them, if you are able, for that belief (which may be called one of the parables of the devil's gospel) that a man rescued from the sea will prove the bane of his deliverer. It might be thought that my grandfather, coming there unknown, and upon an employment so hateful to the inhabitants, must have run the hazard of his life. But this were to misunderstand. He came franked by the laird and the clergyman; he was the King's officer; the work was 'opened with prayer by the Rev. Walter Trail, minister of the parish'; God and the King had decided it, and the people of these pious islands bowed their heads. There landed, indeed, in North Ronaldsay, during the last decade of the eighteenth century, a traveller whose life seems really to have been imperilled. A very little man of a swarthy complexion, he came ashore, exhausted and unshaved, from a long boat passage, and lay down to sleep in the home of the parish schoolmaster. But he had been seen landing. The inhabitants had identified him for a Pict, as, by some singular confusion of name, they called the dark and dwarfish aboriginal people of the land. Immediately the obscure ferment of a race-hatred, grown into a superstition, began to work in their bosoms, and they crowded about the house and the room-door with fearful whisperings. For some time the schoolmaster held them at bay, and at last despatched a messenger to call my grandfather. He came: he found the islanders beside themselves at this unwelcome resurrection of the dead and the detested; he was shown, as adminicular of testimony, the traveller's uncouth and thick-soled boots; he argued, and finding argument unavailing, consented to enter the room and examine with his own eyes the sleeping Pict. One glance was sufficient: the man was now a missionary, but he had been before that an

Edinburgh shopkeeper with whom my grandfather had dealt. He came forth again with this report, and the folk of the island, wholly relieved, dispersed to their own houses. They were timid as sheep and ignorant as limpets; that was all. But the Lord deliver us from the tender mercies of a frightened flock!

I will give two more instances of their superstition. When Sir Walter Scott visited the Stones of Stennis, my grandfather put in his pocket a hundred-foot line, which he unfortunately lost.

'Some years afterwards,' he writes, 'one of my assistants on a visit to the Stones of Stennis took shelter from a storm in a cottage close by the lake; and seeing a box-measuring-line in the bole or sole of the cottage window, he asked the woman where she got this well-known professional appendage. She said: "O sir, ane of the bairns fand it lang syne at the Stanes; and when drawing it out we took fright, and thinking it had belanged to the fairies, we threw it into the bole, and it has layen there ever since."'

This is for the one; the last shall be a sketch by the master hand of Scott himself:

'At the village of Stromness, on the Orkney main island, called Pomona, lived, in 1814, an aged dame called Bessie Millie, who helped out her subsistence by selling favourable winds to mariners. He was a venturous master of a vessel who left the roadstead of Stromness without paying his offering to propitiate Bessie Millie! Her fee was extremely moderate, being exactly sixpence, for which she boiled her kettle and gave the bark the advantage of her prayers, for she disclaimed all unlawful acts. The wind thus petitioned for was sure, she said, to arrive, though occasionally the mariners had to wait some time for it. The woman's dwelling and appearance were not unbecoming her pretensions. Her house, which was on the brow of the steep hill on which Stromness is founded, was only accessible by a series of dirty and precipitous lanes, and for exposure might have been the abode of Eolus himself, in whose commodities the inhabitant dealt. She herself was, as she told us, nearly one hundred years old, withered and dried up like a mummy. A clay-coloured kerchief, folded round her neck, corresponded in colour to her corpse-like complexion. Two light blue eyes that gleamed with a lustre like that of insanity, an utterance of astonishing rapidity, a nose and chin that almost met together, and a ghastly expression of cunning, gave her the effect of Hecate. Such was Bessie Millie, to whom the mariners paid a sort of tribute with a feeling between jest and earnest.'

II

From about the beginning of the century up to 1807 Robert Stevenson was in partnership with Thomas Smith. In the last-named year the partnership was dissolved; Thomas Smith returning to his business, and my grandfather becoming sole engineer to the Board of Northern Lights.

I must try, by excerpts from his diary and correspondence, to convey to the reader some idea of the ardency and thoroughness with which he threw himself into the largest and least of his multifarious engagements in this service. But first I must say a word or two upon the life of lightkeepers, and the temptations to which they are more particularly exposed. The lightkeeper occupies a position apart among men. In sea-towers the complement has always been three since the deplorable business in the Eddystone, when one keeper died, and the survivor, signalling in vain for relief, was compelled to live for days with the dead body. These usually pass their time by the pleasant human expedient of quarrelling; and sometimes, I am assured, not one of the three is on speaking terms with any other. On shore stations, which on the Scottish coast are sometimes hardly less isolated, the usual number is two, a principal and an assistant. The principal is dissatisfied with the assistant, or perhaps the assistant keeps pigeons, and the principal wants the water from the roof. Their wives and families are with them, living cheek by jowl. The children quarrel; Jockie hits Jimsie in the eye, and the mothers make haste to mingle in the dissension. Perhaps there is trouble about a broken dish; perhaps Mrs. Assistant is more highly born than Mrs. Principal and gives herself airs; and the men are drawn in and the servants presently follow. 'Church privileges have been denied the keeper's and the assistant's servants,' I read in one case, and the eminently Scots periphrasis means neither more nor less than excommunication, 'on account of the discordant and quarrelsome state of the families. The cause, when inquired into, proves to be *tittle-tattle* on both sides.' The tender comes round; the foremen and artificers go from station to station; the gossip flies through the whole system of the service, and the stories, disfigured and exaggerated, return to their own birthplace with the returning tender. The English Board was apparently shocked by the picture of these dissensions. 'When the Trinity House can,' I find my grandfather writing at Beachy Head, in 1834, 'they do not appoint two keepers, they disagree so ill. A man who has a family is assisted by his family; and in this way, to my experience and present observation, the business is very much neglected. One keeper is, in my view, a bad system. This day's visit to an English lighthouse convinces me of this, as the lightkeeper was walking on a staff with the gout, and the business performed by one of his daughters, a girl of thirteen or fourteen years of age.' This man received a hundred a year! It shows a different reading of human nature, perhaps typical of Scotland and England, that I find in my grandfather's diary the following pregnant entry: '*The lightkeepers, agreeing ill, keep one another to their duty.*' But the Scottish system was not alone founded on this cynical opinion. The dignity and the comfort of the northern lightkeeper were both attended to. He had a uniform to 'raise him in his own estimation, and in that of his neighbour, which is of consequence to a person of trust. The keepers,' my grandfather goes on, in another place, 'are attended to in all the detail of accommodation in the best style as shipmasters; and this is believed to have a sensible effect upon their conduct, and to regulate their general habits as members of society.' He notes, with the same dip of ink, that 'the brasses were not clean, and the persons of the keepers not *trig*'; and thus we find him writing to a culprit: 'I have to complain that you are not cleanly in your person, and that your manner of speech is ungentle, and rather inclines to rudeness. You must therefore take a different view of your duties as a lightkeeper.' A high ideal for the service appears in these expressions, and will be more amply illustrated further on. But even the Scottish lightkeeper was frail. During the unbroken solitude of the winter months, when inspection is scarce possible, it must seem a vain toil to polish the brass hand-rail of the stair, or to keep an unrewarded vigil in the light-room; and the keepers are habitually tempted to the beginnings of sloth, and must unremittingly resist. He who temporises with his conscience is already lost. I must tell here an anecdote that illustrates the difficulties of inspection. In the days of my uncle David and my father there was a station which they regarded with jealousy. The two engineers compared notes and were agreed. The tower was always clean, but seemed always to bear traces of a hasty cleansing, as though the keepers had been suddenly forewarned. On inquiry, it proved that such was the case, and that a wandering fiddler was the unfailing harbinger of the engineer. At last my father was storm-stayed one Sunday in a port at the other side of the island. The visit was quite overdue, and as he walked across upon the Monday morning he promised himself that he should at last take the keepers unprepared. They were both waiting for him in uniform at the gate; the fiddler had been there on Saturday!

My grandfather, as will appear from the following extracts, was much a martinet, and had a habit of expressing himself on paper with an almost startling emphasis. Personally, with his powerful voice, sanguine countenance, and eccentric and original locutions, he was well qualified to inspire a salutary terror in the service.

'I find that the keepers have, by some means or another, got into the way of cleaning too much with rotten-stone and oil. I take the principal keeper to *task* on this subject, and make him bring a clean towel and clean one of the brazen frames, which leaves the towel in an odious state. This towel I put up in a sheet of paper, seal, and take with me to confront Mr. Murdoch, who has just left the station.' 'This letter' —a stern enumeration of complaints—'to lie a week on the light-room book-place, and to be put in the Inspector's hands when he comes round.' 'It is the most painful thing that can occur for me to have a correspondence of this kind with any of the keepers; and when I come to the Lighthouse, instead of having the satisfaction to meet them with approbation, it is distressing when one is obliged to put on a most angry countenance and demeanour; but from such culpable negligence as you have shown there is no avoiding it. I hold it as a fixed maxim that, when a man or a family put on a slovenly appearance in their houses, stairs, and lanterns, I always find their reflectors, burners, windows, and light in general, ill attended to; and, therefore, I must insist on cleanliness throughout.' 'I find you very deficient in the duty of the high tower. You thus place your appointment as Principal Keeper in jeopardy; and I think it necessary, as an old servant of the Board, to put you upon your guard once for all at this time. I call upon you to recollect what was formerly and is now said to you. The state of the backs of the reflectors at the high tower was disgraceful, as I pointed out to you on the spot. They were as if spitten upon, and greasy finger-marks upon the back straps. I demand an explanation of this state of things.' 'The cause of the Commissioners dismissing you is expressed in the minute; and it must be a matter of regret to you that you have been so much engaged in smuggling, and also that the Reports relative to the cleanliness of the Lighthouse, upon being referred to, rather added to their unfavourable opinion.' 'I do not go into the dwelling-house, but severely chide the lightkeepers for the disagreement that seems to subsist among them.' 'The families of the two lightkeepers here agree very ill. I have effected a reconciliation for the present.' 'Things are in a very *humdrum* state here. There is no painting, and in and out of doors no taste or tidiness displayed. Robert's wife *greets* and M'Gregor's scolds; and Robert is so down-hearted that he says he is unfit for duty. I told him that if he was to mind wives' quarrels, and to take them up, the only way was for him and M'Gregor to go down to the point like Sir G. Grant and Lord Somerset.' 'I cannot say that I have experienced a more unpleasant meeting than that of the lighthouse folks this morning, or ever saw a stronger example of unfeeling barbarity than the conduct which the ——s exhibited. These two cold-hearted persons, not contented with having driven the daughter of the poor nervous woman from her father's house, *both* kept *pouncing* at her, lest she should forget her great misfortune. Write me of their conduct. Do not make any communication of the state of these families at Kinnaird Head, as this would be like *Tale-bearing*.'

There is the great word out. Tales and Tale-bearing, always with the emphatic capitals, run continually in his correspondence. I will give but two instances:—

'Write to David [one of the lightkeepers] and caution him to be more prudent how he expresses himself. Let him attend his duty to the Lighthouse and his family concerns, and give less heed to Tale-bearers.' 'I have not your last letter at hand to quote its date; but, if I recollect, it contains some kind of tales, which nonsense I wish you would lay aside, and notice only the concerns of your family and the important charge committed to you.'

Apparently, however, my grandfather was not himself inaccessible to the Tale-bearer, as the following indicates:

'In walking along with Mr. ——, I explain to him that I should be under the necessity of looking more closely into the business here from his conduct at Buddonness, which had given an instance of weakness in the Moral principle which had staggered my opinion of him. His answer was, "That will be with regard to the lass?" I told him I was to enter no farther with him upon the subject.' 'Mr. Miller appears to be master and man. I am sorry about this foolish fellow. Had I known his train, I should not, as I did, have rather forced him into the service. Upon finding the windows in the state they were, I turned upon Mr. Watt, and especially upon Mr. Stewart. The latter did not appear for a length of time to have visited the light-room. On asking the cause—did Mr. Watt and him (*sic*) disagree; he said no; but he had got very bad usage from the assistant, "who was a very obstreperous man." I could not bring Mr. Watt to put in language his objections to Miller; all I could get was that, he being your friend, and saying he was unwell, he did not like to complain or to push the man; that the man seemed to have no liking to anything like work; that he was unruly; that, being an educated man, he despised them. I was, however, determined to have out of these *unwilling* witnesses the language alluded to. I fixed upon Mr. Stewart as chief; he hedged. My curiosity increased, and I urged. Then he said, "What would I think, just exactly, of Mr. Watt being called an Old B——?" You may judge of my surprise. There was not another word uttered. This was quite enough, as coming from a person I should have calculated upon quite different behaviour from. It spoke a volume of the man's mind and want of principle.' 'Object to the keeper keeping a Bull-Terrier dog of ferocious appearance. It is dangerous, as we land at all times of the night.' 'Have only to complain of the storehouse floor being spotted with oil. Give orders for this being instantly rectified, so that on my return to-morrow I may see things in good order.' 'The furniture of both houses wants much rubbing. Mrs. ——'s carpets are absurd beyond anything I have seen. I want her to turn the fenders up with the bottom to the fireplace: the carpets, when not likely to be in use, folded up and laid as a hearthrug partly under the fender.'

My grandfather was king in the service to his finger-tips. All should go in his way, from the principal lightkeeper's coat to the assistant's fender, from the gravel in the garden-walks to the bad smell in the kitchen, or the oil-spots on the storeroom floor. It might be thought there was nothing more calculated to awake men's resentment, and yet his rule was not more thorough than it was beneficent. His thought for the keepers was continual, and it did not end with their lives. He tried to manage their successions; he thought no pains too great to arrange between a widow and a son who had succeeded his father; he was often harassed and perplexed by tales of hardship; and I find him writing, almost in despair, of their improvident habits and the destitution that awaited their families upon a death. 'The house being completely furnished, they come into possession without necessaries, and they go out NAKED. The insurance seems to have failed, and what next is to be tried?' While they lived he wrote behind their backs to arrange for the education of their children, or to get them other situations if they seemed unsuitable for the Northern Lights. When he was at a lighthouse on a Sunday he held prayers and heard the children read. When a keeper was sick, he lent him his horse and sent him mutton and brandy from the ship. 'The assistant's wife having been this morning confined, there was sent ashore a bottle of sherry and a few rusks—a practice which I have always observed in this service,' he writes. They dwelt, many of them, in uninhabited isles or desert forelands, totally cut off from shops Many of them were, besides, fallen into a rustic dishabitude of life, so that even when they visited a city they could scarce be trusted with their own affairs, as (for example) he who carried home to his children, thinking they were oranges, a bag of lemons. And my grandfather seems to have acted, at least in his early years, as a kind of gratuitous agent for the service. Thus I find him writing to a keeper in 1806, when his mind was already preoccupied with arrangements for the Bell Rock: 'I am much afraid I stand very unfavourably with you as a man of promise, as I was to send several things of which I believe I have more than once got the memorandum. All I can say is that in this respect you are not singular. This makes me no better; but really I have been driven about beyond all example in my past experience, and have been essentially obliged to neglect my own urgent affairs.' No servant of the Northern Lights came to Edinburgh but he was entertained at Baxter's Place to breakfast. There, at his own table, my grandfather sat down delightedly with his broad-spoken, homespun officers. His whole relation to the service was, in fact, patriarchal; and I believe I may say that throughout its ranks he was adored. I have spoken with many who knew him; I was his grandson, and their words may have very well been words of flattery; but there was one thing that could not be affected, and that was the look and light that came into their faces at the name of Robert Stevenson.

In the early part of the century the foreman builder was a young man of the name of George Peebles, a native of Anstruther. My grandfather had placed in him a very high degree of confidence, and he was already designated to be foreman at the Bell Rock, when, on Christmas-day 1806, on his way home from Orkney, he was lost in the schooner *Traveller*. The tale of the loss of the *Traveller* is almost a replica of that of the *Elizabeth* of Stromness; like the *Elizabeth* she came as far as Kinnaird Head, was then surprised by a storm, driven back to Orkney, and bilged and sank on the island of Flotta. It seems it was about the dusk of the day when the ship struck, and many of the crew and passengers were drowned. About the same hour, my grandfather was in his office at the writing-table; and the room beginning to darken, he laid down his pen and fell asleep. In a dream he saw the door open and George Peebles come in, 'reeling to and fro, and staggering like a drunken man,' with water streaming from his head and body to the floor. There it gathered into a wave which, sweeping forward, submerged my grandfather. Well, no matter how deep; versions vary; and at last he awoke, and behold it was a dream! But it may be conceived how profoundly the impression was written even on the mind of a man averse from such ideas, when the news came of the wreck on Flotta and the death of George.

George's vouchers and accounts had perished with himself; and it appeared he was in debt to the Commissioners. But my grandfather wrote to Orkney twice, collected evidence of his disbursements, and proved him to be seventy pounds ahead. With this sum, he applied to George's brothers, and had it apportioned between their mother and themselves. He approached the Board and got an annuity of £5 bestowed on the widow Peebles; and we find him writing her a long letter of explanation and advice, and pressing on her the duty of making a will. That he should thus act executor was no singular instance. But besides this we are able to assist at some of the stages of a rather touching experiment; no less than an attempt to secure Charles Peebles heir to George's favour. He is despatched, under the character of 'a fine young man'; recommended to gentlemen for 'advice, as he's a stranger in your place, and indeed to this kind of charge, this being his first outset as Foreman'; and for a long while after, the letter-book, in the midst of that thrilling first year of the Bell Rock, is encumbered with pages of instruction and encouragement. The nature of a bill, and the precautions that are to be observed about discounting it, are expounded at length and with clearness. 'You are not, I hope, neglecting, Charles, to work the harbour at spring-tides; and see that you pay the greatest attention to get the well so as to supply the keeper with water, for he is a very helpless fellow, and so unfond of hard work that I fear he could do ill to keep himself in water by going to the other side for it.'—'With regard to spirits, Charles, I see very little occasion for it.' These abrupt apostrophes sound to me like the voice of an awakened conscience; but they would seem to have reverberated in vain in the ears of Charles. There was trouble in Pladda, his scene of operations; his men ran away from him, there was at least a talk of calling in the Sheriff. 'I fear,' writes my grandfather, 'you have been too indulgent, and I am sorry to add that men do not answer to be too well treated, a circumstance which I have experienced, and which you will learn as you go on in business.' I wonder, was not Charles Peebles himself a case in point? Either death, at least, or disappointment and discharge, must have ended his service in the Northern Lights; and in later correspondence I look in vain for any mention of his name—Charles, I mean, not Peebles: for as late as 1839 my grandfather is patiently writing to another of the family: 'I am sorry you took the trouble of applying to me about your son, as it lies quite out of my way to forward his views in the line of his profession as a Draper.'

III

A professional life of Robert Stevenson has been already given to the world by his son David, and to that I would refer those interested in such matters. But my own design, which is to represent the man, would be very ill carried out if I suffered myself or my reader to forget that he was, first of all and last of all, an engineer. His chief claim to the style of a mechanical inventor is on account of the Jib or Balance Crane of the Bell Rock, which are beautiful contrivances. But the great merit of this engineer was not in the field of engines. He was above all things a projector of works in the face of nature, and a modifier of nature itself. A road to be made, a tower to be built, a harbour to be constructed, a river to be trained and guided in its channel—these were the problems with which his mind was continually occupied; and for these and similar ends he travelled the world for more than half a century, like an artist, note-book in hand.

He once stood and looked on at the emptying of a certain oil-tube; he did so watch in hand, and accurately timed the operation; and in so doing offered the perfect type of his profession. The fact acquired might never be of use: it was acquired: another link in the world's huge chain of processes was brought down to figures and placed at the service of the engineer. 'The very term mensuration sounds *engineer-like*,' I find him writing; and in truth what the engineer most properly deals with is that which can be measured, weighed, and numbered. The time of any operation in hours and minutes, its cost in pounds, shillings, and pence, the strain upon a given point in foot-pounds—these are his conquests, with which he must continually furnish his mind, and which, after he has acquired them, he must continually apply and exercise. They must be not only entries in note-books, to be hurriedly consulted; in the actor's phrase, he must be *stale* in them; in a word of my grandfather's, they must be 'fixed in the mind like the ten fingers and ten toes.'

These are the certainties of the engineer; so far he finds a solid footing and clear views. But the province of formulas and constants is restricted. Even the mechanical engineer comes at last to an end of his figures, and must stand up, a practical man, face to face with the discrepancies of nature and the hiatuses of theory. After the machine is finished, and the steam turned on, the next is to drive it; and experience and an exquisite sympathy must teach him where a weight should be applied or a nut loosened. With the civil engineer, more properly so called (if anything can be proper with this awkward coinage), the obligation starts with the beginning. He is always the practical man. The rains, the winds and the waves, the complexity and the fitfulness of nature, are always before him. He has to deal with the unpredictable, with those forces (in Smeaton's phrase) that 'are subject to no calculation'; and still he must predict, still calculate them, at his peril. His work is not yet in being, and he must foresee its influence: how it shall deflect the tide, exaggerate the waves, dam back the rain-water, or attract the thunderbolt. He visits a piece of sea-board; and from the inclination and soil of the beach, from the weeds and shell-fish, from the configuration of the coast and the depth of soundings outside, he must deduce what magnitude of waves is to be looked for. He visits a river, its summer water babbling on shallows; and he must not only read, in a thousand indications, the measure of winter freshets, but be able to predict the violence of occasional great floods. Nay, and more; he must not only consider that which is, but that which may be. Thus I find my grandfather writing, in a report on the North Esk Bridge: 'A less waterway might have sufficed, but *the valleys may come to be meliorated by drainage.*' One field drained after another through all that confluence of vales, and we come to a time when they shall precipitate by so much a more copious and transient flood, as the gush of the flowing drain-pipe is superior to the leakage of a peat.

It is plain there is here but a restricted use for formulas. In this sort of practice, the engineer has need of some transcendental sense. Smeaton, the pioneer, bade him obey his 'feelings'; my father, that 'power of estimating obscure forces which supplies a coefficient of its own to every rule.' The rules must be everywhere indeed; but they must everywhere be modified by this transcendental coefficient, everywhere bent to the impression of the trained eye and the *feelings* of the engineer. A sentiment of physical laws and of the scale of nature, which shall have been strong in the beginning and progressively fortified by observation, must be his guide in the last recourse. I had the most opportunity to observe my father. He would pass hours on the beach, brooding over the waves, counting them, noting their least deflection, noting when they broke. On Tweedside, or by Lyne or Manor, we have spent together whole afternoons; to me, at the time, extremely wearisome; to him, as I am now sorry to think, bitterly mortifying. The river was to me a pretty and various spectacle; I could not see—I could not be made to see—it otherwise. To my father it was a chequer-board of lively forces, which he traced from pool to shallow with minute appreciation and enduring interest. 'That bank was being undercut,' he might say. 'Why? Suppose you were to put a groin out here, would not the *filum fluminis* be cast abruptly off across the channel? and where would it impinge upon the other shore? and what would be the result? Or suppose you were to blast that boulder, what would happen? Follow it—use the eyes God has given you—can you not see that a great deal of land would be reclaimed upon this side?' It was to me like school in holidays; but to him, until I had worn him out with my invincible triviality, a delight. Thus he pored over the engineer's voluminous handy-book of nature; thus must, too, have pored my grandfather and uncles.

But it is of the essence of this knowledge, or this knack of mind, to be largely incommunicable. 'It cannot be imparted to another,' says my father. The verbal casting-net is thrown in vain over these evanescent, inferential relations. Hence the insignificance of much engineering literature. So far as the science can be reduced to formulas or diagrams, the book is to the point; so far as the art depends on intimate study of the ways of nature, the author's words will too often be found vapid. This fact—that engineering looks one way, and literature another—was what my grandfather overlooked. All his life long, his pen was in his hand, piling up a treasury of knowledge, preparing himself against all possible contingencies. Scarce anything fell under his notice but he perceived in it some relation to his work, and chronicled it in the pages of his journal in his always lucid, but sometimes inexact and wordy, style. The Travelling Diary (so he called it) was kept in fascicles of ruled paper, which were at last bound up, rudely indexed, and put by for future reference. Such volumes as have reached me contain a surprising medley: the whole details of his employment in the Northern Lights and his general practice; the whole biography of an enthusiastic engineer. Much of it is useful and curious; much merely otiose; and much can only be described as an attempt to impart that which cannot be imparted in words. Of such are his repeated and heroic descriptions of reefs; monuments of misdirected literary energy, which leave upon the mind of the reader no effect but that of a multiplicity of words and the suggested vignette of a lusty old gentleman scrambling among tangle. It is to be remembered that he came to engineering while yet it was in the egg and without a library, and that he saw the bounds of that profession widen daily. He saw iron ships, steamers, and the locomotive engine, introduced. He lived to travel from Glasgow to Edinburgh in the inside of a forenoon, and to remember that he himself had 'often been twelve hours upon the journey, and his grandfather (Lillie) two days'! The profession was still but in its second generation, and had already broken down the barriers of time and space. Who should set a limit to its future encroachments? And hence, with a kind of sanguine pedantry, he pursued his design of 'keeping up with the day' and posting himself and his family on every mortal subject. Of this unpractical idealism we shall meet with many instances; there was not a trade, and scarce an accomplishment, but he thought it should form part of the outfit of an engineer; and not content with keeping an encyclopædic diary himself, he would fain have set all his sons to work continuing and extending it. They were more happily inspired. My father's engineering pocket-book was not a bulky volume; with its store of pregnant notes and vital formulas, it served him through life, and was not yet filled when he came to die. As for Robert Stevenson and the Travelling Diary, I should be ungrateful to complain, for it has supplied me with many lively traits for this and subsequent chapters; but I must still remember much of the period of my study there as a sojourn in the Valley of the Shadow.

The duty of the engineer is twofold—to design the work, and to see the work done. We have seen already something of the vociferous thoroughness of the man, upon the cleaning of lamps and the polishing of reflectors. In building, in road-making, in the construction of bridges, in every detail and byway of his employments, he pursued the same ideal. Perfection (with a capital P and violently under-scored) was his design. A crack for a penknife, the waste of 'six-and-thirty shillings,' 'the loss of a day or a tide,' in each of these he saw and was revolted by the finger of the sloven; and to spirits intense as his, and immersed in vital undertakings, the slovenly is the dishonest, and wasted time is instantly translated into lives endangered. On this consistent idealism there is but one thing that now and then trenches with a touch of incongruity, and that is his love of the picturesque. As when he laid out a road on Hogarth's line of beauty; bade a foreman be careful, in quarrying, not 'to disfigure the island'; or regretted in a report that 'the great stone, called the *Devil in the Hole*, was blasted or broken down to make road-metal, and for other purposes of the work.'

CHAPTER III

THE BUILDING OF THE BELL ROCK

Off the mouths of the Tay and the Forth, thirteen miles from Fifeness, eleven from Arbroath, and fourteen from the Red Head of Angus, lies the Inchcape or Bell Rock. It extends to a length of about fourteen hundred feet, but the part of it discovered at low water to not more than four hundred and twenty-seven. At a little more than half-flood in fine weather the seamless ocean joins over the reef, and at high-water springs it is buried sixteen feet. As the tide goes down, the higher reaches of the rock are seen to be clothed by *Conferva rupestris* as by a sward of grass; upon the more exposed edges, where the currents are most swift and the breach of the sea heaviest, Baderlock or Henware flourishes; and the great Tangle grows at the depth of several fathoms with luxuriance. Before man arrived, and introduced into the silence of the sea the smoke and clangour of a blacksmith's shop, it was a favourite resting-place of seals. The crab and lobster haunt in the crevices; and limpets, mussels, and the white buckie abound.

According to a tradition, a bell had been once hung upon this rock by an abbot of Arbroath,[1] 'and being taken down by a sea-pirate, a year thereafter he perished upon the same rock, with ship and goods, in the righteous judgment of God.' From the days of the abbot and the sea-pirate no man had set foot upon the Inchcape, save fishers from the neighbouring coast, or perhaps—for a moment, before the surges swallowed them—the unfortunate victims of shipwreck. The fishers approached the rock with an extreme timidity; but their harvest appears to have been great, and the adventure no more perilous than lucrative. In 1800, on the occasion of my grandfather's first landing, and during the two or three hours which the ebb-tide and the smooth water allowed them to pass upon its shelves, his crew collected upwards of two hundredweight of old metal: pieces of a kedge anchor and a cabin stove, crowbars, a hinge and lock of a door, a ship's marking-iron, a piece of a ship's caboose, a soldier's bayonet, a cannon ball, several pieces of money, a shoe-buckle, and the like. Such were the spoils of the Bell Rock.

[1] This is, of course, the tradition commemorated by Southey in his ballad of 'The Inchcape Bell.' Whether true or not, it points to the fact that from the infancy of Scottish navigation, the seafaring mind had been fully alive to the perils of this reef. Repeated attempts had been made to mark the place with beacons, but all efforts were unavailing (one such beacon having been carried away within eight days of its erection) until Robert Stevenson conceived and carried out the idea of the stone tower.

But the number of vessels actually lost upon the reef was as nothing to those that were cast away in fruitless efforts to avoid it. Placed right in the fairway of two navigations, and one of these the entrance to the only harbour of refuge between the Downs and the Moray Firth, it breathed abroad along the whole coast an atmosphere of terror and perplexity; and no ship sailed that part of the North Sea at night, but what the ears of those on board would be strained to catch the roaring of the seas on the Bell Rock.

From 1794 onward, the mind of my grandfather had been exercised with the idea of a light upon this formidable danger. To build a tower on a sea rock, eleven miles from shore, and barely uncovered at low water of neaps, appeared a fascinating enterprise. It was something yet unattempted, unessayed; and even now, after it has been lighted for more than eighty years, it is still an exploit that has never been repeated.[1]

[1] The particular event which concentrated Mr. Stevenson's attention on the problem of the Bell Rock was the memorable gale of December 1799, when, among many other vessels, H.M.S. *York*, a seventy-four-gun ship, went down with all hands on board. Shortly after this disaster Mr. Stevenson made a careful survey, and prepared his models for a stone tower, the idea of which was at first received with pretty general scepticism. Smeaton's Eddystone tower could not be cited as affording a parallel, for there the rock is not submerged even at high-water, while the problem of the Bell Rock was to build a tower of masonry on a sunken reef far distant from land, covered at every tide to a depth of twelve feet or more, and having thirty-two fathoms' depth of water within a mile of its eastern edge.

My grandfather was, besides, but a young man, of an experience comparatively restricted, and a reputation confined to Scotland; and when he prepared his first models, and exhibited them in Merchants' Hall, he can hardly be acquitted of audacity. John Clerk of Eldin stood his friend from the beginning, kept the key of the model room, to which he carried 'eminent strangers,' and found words of counsel and encouragement beyond price. 'Mr. Clerk had been personally known to Smeaton, and used occasionally to speak of him to me,' says my grandfather; and again: 'I felt regret that I had not the opportunity of a greater range of practice to fit me for such an undertaking; but I was fortified by an expression of my friend Mr. Clerk in one of our conversations. "This work," said he, "is unique, and can be little forwarded by experience of ordinary masonic operations. In this case Smeaton's 'Narrative' must be the text-book, and energy and perseverance the pratique."'

A Bill for the work was introduced into Parliament and lost in the Lords in 1802-3. John Rennie was afterwards, at my grandfather's suggestion, called in council, with the style of chief engineer. The precise meaning attached to these words by any of the parties appears irrecoverable. Chief engineer should have full authority, full responsibility, and a proper share of the emoluments;

and there were none of these for Rennie. I find in an appendix a paper which resumes the controversy on this subject; and it will be enough to say here that Rennie did not design the Bell Rock, that he did not execute it, and that he was not paid for it.[1] From so much of the correspondence as has come down to me, the acquaintance of this man, eleven years his senior, and already famous, appears to have been both useful and agreeable to Robert Stevenson. It is amusing to

[1] The grounds for the rejection of the Bill by the House of Lords in 1802-3 had been that the extent of coast over which dues were proposed to be levied would be too great. Before going to Parliament again, the Board of Northern Lights, desiring to obtain support and corroboration for Mr. Stevenson's views, consulted first Telford, who was unable to give the matter his attention, and then (on Stevenson's suggestion) Rennie, who concurred in affirming the practicability of a stone tower, and supported the Bill when it came again before Parliament in 1806. Rennie was afterwards appointed by the Commissioners as advising engineer, whom Stevenson might consult in cases of emergency. It seems certain that the title of chief engineer had in this instance no more meaning than the above. Rennie, in point of fact, proposed certain modifications in Stevenson's plans, which the latter did not accept; nevertheless Rennie continued to take a kindly interest in the work, and the two engineers remained in friendly correspondence during its progress. The official view taken by the Board as to the quarter in which lay both the merit and the responsibility of the work may be gathered from a minute of the Commissioners at their first meeting held after Stevenson died; in which they record their regret 'at the death of this zealous, faithful, and able officer, *to whom is due the honour of conceiving and executing the Bell Rock Lighthouse.*' The matter is briefly summed up in the *Life* of Robert Stevenson by his son David Stevenson (A. & C. Black, 1878), and fully discussed, on the basis of official facts and figures, by the same writer in a letter to the *Civil Engineers' and Architects' Journal*, 1862.

find my grandfather seeking high and low for a brace of pistols which his colleague had lost by the way between Aberdeen and Edinburgh; and writing to Messrs. Dollond, 'I have not thought it necessary to trouble Mr. Rennie with this order, but *I beg you will see to get two minutes of him as he passes your door*'—a proposal calculated rather from the latitude of Edinburgh than from London, even in 1807. It is pretty, too, to observe with what affectionate regard Smeaton was held in mind by his immediate successors. 'Poor old fellow,' writes Rennie to Stevenson, 'I hope he will now and then take a peep at us, and inspire you with fortitude and courage to brave all difficulties and dangers to accomplish a work which will, if successful, immortalise you in the annals of fame.' The style might be bettered, but the sentiment is charming.

Smeaton was, indeed, the patron saint of the Bell Rock. Undeterred by the sinister fate of Winstanley, he had tackled and solved the problem of the Eddystone; but his solution had not been in all respects perfect. It remained for my grandfather to outdo him in daring, by applying to a tidal rock those principles which had been already justified by the success of the Eddystone, and to perfect the model by more than one exemplary departure. Smeaton had adopted in his floors the principle of the arch; each therefore exercised an outward thrust upon the walls, which must be met and combated by embedded chains. My grandfather's flooring-stones, on the other hand, were flat, made part of the outer wall, and were keyed and dovetailed into a central stone, so as to bind the work together and be positive elements of strength. In 1703 Winstanley still thought it possible to erect his strange pagoda, with its open gallery, its florid scrolls and candlesticks: like a rich man's folly for an ornamental water in a park. Smeaton followed; then Stevenson in his turn corrected such flaws as were left in Smeaton's design; and with his improvements, it is not too much to say the model was made perfect. Smeaton and Stevenson had between them evolved and finished the sea-tower. No subsequent builder has departed in anything essential from the principles of their design. It remains, and it seems to us as though it must remain for ever, an ideal attained. Every stone in the building, it may interest the reader to know, my grandfather had himself cut out in the model; and the manner in which the courses were fitted, joggled, trenailed, wedged, and the bond broken, is intricate as a puzzle and beautiful by ingenuity.

In 1806 a second Bill passed both Houses, and the preliminary works were at once begun. The same year the Navy had taken a great harvest of prizes in the North Sea, one of which, a Prussian fishing dogger, flat-bottomed and rounded at the stem and stern, was purchased to be a floating lightship, and re-named the *Pharos*. By July 1807 she was overhauled, rigged for her new purpose, and turned into the lee of the Isle of May. 'It was proposed that the whole party should meet in her and pass the night; but she rolled from side to side in so extraordinary a manner, that even the most seahardy fled. It was humorously observed of this vessel that she was in danger of making a round turn and appearing with her keel uppermost; and that she would even turn a halfpenny if laid upon deck.' By two o'clock on the morning of the 15th July this purgatorial vessel was moored by the Bell Rock.

A sloop of forty tons had been in the meantime built at Leith, and named the *Smeaton*; by the 7th of August my grandfather set sail in her—

'carrying with him Mr. Peter Logan, foreman builder, and five artificers selected from their having been somewhat accustomed to the sea, the writer being aware of the distressing trial which the floating light would necessarily inflict upon landsmen from her rolling motion. Here he remained till the 10th, and, as the weather was favourable, a landing was effected daily, when the workmen were employed in cutting the large seaweed from the sites of the lighthouse and beacon, which were respectively traced with pickaxes upon the rock. In the meantime the crew of the *Smeaton* was employed in laying down the several sets of moorings within about half a mile of the rock for the convenience of vessels. The artificers, having, fortunately, experienced moderate weather, returned to the workyard of Arbroath with a good report of their treatment afloat; when their comrades ashore began to feel some anxiety to see a place of which they had heard so much, and to change the constant operations with the iron and mallet in the process of hewing for an occasional tide's work on the rock, which they figured to themselves as a state of comparative ease and comfort.'

I am now for many pages to let my grandfather speak for himself, and tell in his own words the story of his capital achievement. The tall quarto of 533 pages from which the following narrative has been dug out is practically unknown to the general reader, yet good judges have perceived its merit, and it has been named (with flattering wit) 'The Romance of Stone and Lime' and 'The Robinson Crusoe of Civil Engineering. The tower was but four years in the building; it took Robert Stevenson, in the midst of his many avocations, no less than fourteen to prepare the *Account.* The title-page is a solid piece of literature of upwards of a hundred words; the table of contents runs to thirteen pages; and the dedication (to that revered monarch, George IV) must have cost him no little study and correspondence. Walter Scott was called in council, and offered one miscorrection which still blots the page. In spite of all this pondering and filing, there remain pages not easy to construe, and inconsistencies not easy to explain away. I have sought to make these disappear, and to lighten a little the baggage with which my grandfather marches; here and there I have rejointed and rearranged a sentence, always with his own words, and all with a reverent and faithful hand; and I offer here to the reader the true Monument of Robert Stevenson with a little of the moss removed from the inscription, and the Portrait of the artist with some superfluous canvas cut away.

I

OPERATIONS OF 1807

1807

Sunday, 16th Aug.

Everything being arranged for sailing to the rock on Saturday the 15th, the vessel might have proceeded on the Sunday; but understanding that this would not be so agreeable to the artificers it was deferred until Monday. Here we cannot help observing that the men allotted for the operations at the rock seemed to enter upon the undertaking with a degree of consideration which fully marked their opinion as to the hazardous nature of the undertaking on which they were about to enter. They went in a body to church on Sunday, and whether it was in the ordinary course, or designed for the occasion, the writer is not certain, but the service was, in many respects, suitable to their circumstances.

Monday, 17th Aug.

The tide happening to fall late in the evening of Monday the 17th, the party, counting twenty-four in number, embarked on board of the *Smeaton* about ten o'clock p.m., and sailed from Arbroath with a gentle breeze at west. Our ship's colours having been flying all day in compliment to the commencement of the work, the other vessels in the harbour also saluted, which made a very gay appearance. A number of the friends and acquaintances of those on board having been thus collected, the piers, though at a late hour, were perfectly crowded, and just as the *Smeaton* cleared the harbour, all on board united in giving three hearty cheers, which were returned by those on shore in such good earnest, that, in the still of the evening, the sound must have been heard in all parts of the town, re-echoing from the walls and lofty turrets of the venerable Abbey of Aberbrothwick. The writer felt much satisfaction at the manner of this parting scene, though he must own that the present rejoicing was, on his part, mingled with occasional reflections upon the responsibility of his situation, which extended to the safety of all who should be engaged in this perilous work. With such sensations he retired to his cabin; but as the artificers were rather inclined to move about the deck than to remain in their confined berths below, his repose was transient, and the vessel being small every motion was necessarily heard. Some who were musically inclined occasionally sung; but he listened with peculiar pleasure to the sailor at the helm, who hummed over Dibdin's characteristic air:—

1807

> 'They say there 's a Providence sits up aloft,
> To keep watch for the life of poor Jack.'

The weather had been very gentle all night, and, about four in the morning of the 18th, the *Smeaton* anchored. Agreeably to an arranged plan of operations, all hands were called at five o'clock a.m., just as the highest part of the Bell Rock began to show its sable head among the light breakers, which occasionally whitened with the foaming sea. The two boats belonging to the floating light attended the *Smeaton*, to carry the artificers to the rock, as her boat could only accommodate about six or eight sitters. Every one was more eager than his neighbour to leap into the boats, and it required a good deal of management on the part of the coxswains to get men unaccustomed to a boat to take their places for rowing and at the same time trimming her properly. The landing-master and foreman went into one boat, while the writer took charge of another, and steered it to and

Tuesday, 18th Aug.

1807 from the rock. This became the more necessary in the early stages of the work, as places could not be spared for more than two, or at most three, seamen to each boat, who were always stationed, one at the bow, to use the boat-hook in fending or pushing off, and the other at the aftermost oar, to give the proper time in rowing, while the middle oars were double-banked, and rowed by the artificers.

As the weather was extremely fine, with light airs of wind from the east, we landed without difficulty upon the central part of the rock at half-past five, but the water had not yet sufficiently left it for commencing the work. This interval, however, did not pass unoccupied. The first and last of all the principal operations at the Bell Rock were accompanied by three hearty cheers from all hands, and, on occasions like the present, the steward of the ship attended, when each man was regaled with a glass of rum. As the water left the rock about six, some began to bore the holes for the great bats or holdfasts, for fixing the beams of the Beacon-house, while the smith was fully attended in laying out the site of his forge, upon a somewhat sheltered spot of the rock, which also recommended itself from the vicinity of a pool of water for tempering his irons. These preliminary steps occupied about an hour, and as nothing further could be done during this tide towards fixing the forge, the workmen gratified their curiosity by roaming about the rock, which they investigated with great eagerness till the tide overflowed it. Those who had been sick picked dulse (*Fucus palmatus*), which they ate with much seeming appetite; others were more intent upon collecting limpets for bait, to enjoy the amusement of fishing when they returned on board of the vessel. Indeed, none came away empty-handed, as everything found upon the Bell Rock was considered valuable, being connected with some interesting association. Several coins, and numerous bits of shipwrecked iron, were picked up, of almost every description; and, in particular, a marking-iron lettered JAMES—a circumstance of which it was thought proper to give notice to the public, as it might lead to the knowledge of some unfortunate shipwreck, perhaps unheard of till this simple occurrence led to the discovery. When the rock began to be overflowed, the landing-master arranged the crews of the respective boats, appointing twelve persons to each. According to a rule which the writer had laid down to himself, he was always the last person who left the rock.

In a short time the Bell Rock was laid completely under water, and the weather being extremely fine, the sea was so smooth that its place could not be pointed out from the appearance of the surface—a circumstance which sufficiently demonstrates the dangerous nature of this rock, even during the day, and in the smoothest and calmest state of the sea. During the interval between the morning and the evening tides, the artificers were variously employed in fishing and reading; others were busy in drying and adjusting their wet clothes, and one or two amused their companions with the violin and German flute.

About seven in the evening the signal bell for landing on the rock was again rung, when every man was at his quarters. In this service it was thought more appropriate to use the bell than to *pipe* to quarters, as the use of this instrument is less known to the mechanic than the sound of the bell. The landing, as in the morning, was at the eastern harbour. During this tide the seaweed was pretty well cleared

1807 from the site of the operations, and also from the tracks leading to the different landing-places; for walking upon the rugged surface of the Bell Rock, when covered with seaweed, was found to be extremely difficult and even dangerous. Every hand that could possibly be occupied was now employed in assisting the smith to fit up the apparatus for his forge. At 9 p.m. the boats returned to the tender, after other two hours' work, in the same order as formerly—perhaps as much gratified with the success that attended the work of this day as with any other in the whole course of the operations. Although it could not be said that the fatigues of this day had been great, yet all on board retired early to rest. The sea being calm, and no movement on deck, it was pretty generally remarked in the morning that the bell awakened the greater number on board from their first sleep; and though this observation was not altogether applicable to the writer himself, yet he was not a little pleased to find that thirty people could all at once become so reconciled to a night's quarters within a few hundred paces of the Bell Rock.

Wednesday, 19th Aug.

Being extremely anxious at this time to get forward with fixing the smith's forge, on which the progress of the work at present depended, the writer requested that he might be called at daybreak to learn the landing-master's opinion of the weather from the appearance of the rising sun, a criterion by which experienced seamen can generally judge pretty accurately of the state of the weather for the following day. About five o'clock, on coming upon deck, the sun's upper limb or disc had just begun to appear as if rising from the ocean, and in less than a minute he was seen in the fullest splendour; but after a short interval he was enveloped in a soft cloudy sky, which was considered emblematical of fine weather. His rays had not yet sufficiently dispelled the clouds which hid the land from view, and the Bell Rock being still overflowed, the whole was one expanse of water. This scene in itself was highly gratifying; and, when the morning bell was tolled, we were gratified with the happy forebodings of good weather and the expectation of having both a morning and an evening tide's work on the rock.

The boat which the writer steered happened to be the last which approached the rock at this tide; and, in standing up in the stern, while at some distance, to see how the leading boat entered the creek, he was astonished to observe something in the form of a human figure, in a reclining posture, upon one of the ledges of the rock. He immediately steered the boat through a narrow entrance to the eastern harbour, with a thousand unpleasant sensations in his mind. He thought a vessel or boat must have been wrecked upon the rock during the night; and it seemed probable that the rock might be strewed with dead bodies, a spectacle which could not fail to deter the artificers from returning so freely to their work. In the midst of these reveries the boat took the ground at an improper landing-place; but, without waiting to push her off, he leapt upon the rock, and making his way hastily to the spot which had privately given him alarm, he had the satisfaction to ascertain that he had only been deceived by the peculiar situation and aspect of the smith's anvil and block, which very completely represented the appearance of a lifeless body upon the rock. The writer carefully suppressed his feelings, the simple mention of which might have had a bad effect upon the artificers, and his haste passed for an anxiety to examine the apparatus of the

1807 smith's forge, left in an unfinished state at evening tide.

In the course of this morning's work two or three apparently distant peals of thunder were heard, and the atmosphere suddenly became thick and foggy. But as the *Smeaton*, our present tender, was moored at no great distance from the rock, the crew on board continued blowing with a horn, and occasionally fired a musket, so that the boats got to the ship without difficulty.

Thursday, 20th Aug. The wind this morning inclined from the north-east, and the sky had a heavy and cloudy appearance, but the sea was smooth, though there was an undulating motion on the surface, which indicated easterly winds, and occasioned a slight surf upon the rock. But the boats found no difficulty in landing at the western creek at half-past seven, and, after a good tide's work, left it again about a quarter from eleven. In the evening the artificers landed at half-past seven, and continued till half-past eight, having completed the fixing of the smith's forge, his vice, and a wooden board or bench, which were also batted to a ledge of the rock, to the great joy of all, under a salute of three hearty cheers. From an oversight on the part of the smith, who had neglected to bring his tinder-box and matches from the vessel, the work was prevented from being continued for at least an hour longer.

The smith's shop was, of course, in *open space*: the large bellows were carried to and from the rock every tide, for the serviceable condition of which, together with the tinder-box, fuel, and embers of the former fire, the smith was held responsible. Those who have been placed in situations to feel the inconveniency and want of this useful artisan, will be able to appreciate his value in a case like the present. It often happened, to our annoyance and disappointment, in the early state of the work, when the smith was in the middle of a *favourite heat* in making some useful article, or in sharpening the tools, after the flood-tide had obliged the pickmen to strike work, a sea would come rolling over the rocks, dash out the fire, and endanger his indispensable implement, the bellows. If the sea was smooth, while the smith often stood at work knee-deep in water, the tide rose by imperceptible degrees, first cooling the exterior of the fireplace, or hearth, and then quietly blackening and extinguishing the fire from below. The writer has frequently been amused at the perplexing anxiety of the blacksmith when coaxing his fire and endeavouring to avert the effects of the rising tide.

1807

Everything connected with the forge being now completed, the artificers found no want of sharp tools, and the work went forward with great alacrity and spirit. It was also alleged that the rock had a more habitable appearance from the volumes of smoke which ascended from the smith's shop and the busy noise of his anvil, the operations of the masons, the movements of the boats, and shipping at a distance—all contributed to give life and activity to the scene. This noise and traffic had, however, the effect of almost completely banishing the herd of seals which had hitherto frequented the rock as a resting-place during the period of low water. The rock seemed to be peculiarly adapted to their habits, for, excepting two or three days at neap-tides, a part of it always dries at low water—at least, during the summer season—and as there was good fishing-ground in the neighbourhood, without a human being to disturb or molest

Friday, 21st Aug.

1807 them, it had become a very favourite residence of these amphibious animals, the writer having occasionally counted from fifty to sixty playing about the rock at a time. But when they came to be disturbed every tide, and their seclusion was broken in upon by the kindling of great fires, together with the beating of hammers and picks during low water, after hovering about for a time, they changed their place, and seldom more than one or two were to be seen about the rock upon the more detached outlayers which dry partially, whence they seemed to look with that sort of curiosity which is observable in these animals when following a boat.

Saturday, 22nd Aug. Hitherto the artificers had remained on board the *Smeaton*, which was made fast to one of the mooring buoys at a distance only of about a quarter of a mile from the rock, and, of course, a very great conveniency to the work. Being so near, the seamen could never be mistaken as to the progress of the tide, or state of the sea upon the rock, nor could the boats be much at a loss to pull on board of the vessel during fog, or even in very rough weather; as she could be cast loose from her moorings at pleasure, and brought to the lee side of the rock. But the *Smeaton* being only about forty register tons, her accommodations were extremely limited. It may, therefore, be easily imagined that an addition of twenty-four persons to her own crew must have rendered the situation of those on board rather uncomfortable. The only place for the men's hammocks on board being in the hold, they were unavoidably much crowded: and if the weather had required the hatches to be fastened down, so great a number of men could not possibly have been accommodated. To add to this evil, the *co-boose* or cooking-place being upon deck, it would not have been possible to have cooked for so large a company in the event of bad weather.

The stock of water was now getting short, and some necessaries being also wanted for the floating light, the *Smeaton* was despatched for Arbroath; and the writer, with the artificers, at the same time shifted their quarters from her to the floating light.

Although the rock barely made its appearance at this period of the tides till eight o'clock, yet, having now a full mile to row from the floating light to the rock, instead of about a quarter of a mile from the moorings of the *Smeaton*, it was necessary to be earlier astir, and to form different arrangements; breakfast was accordingly served up at seven o'clock this morning. From the excessive motion of the floating light, the writer had looked forward rather with anxiety to the removal of the workmen to this ship. Some among them, who had been congratulating themselves upon having become sea-hardy while on board the *Smeaton*, had a complete relapse upon returning to the floating light. This was the case with the writer. From the spacious and convenient berthage of the floating light, the exchange to the artificers was, in this respect, much for the better. The boats were also commodious, measuring sixteen feet in length on the keel, so that, in fine weather, their complement of sitters was sixteen persons for each, with which, however, they were rather crowded, but she could not stow two boats of larger dimensions. When there was what is called a breeze of wind, and a swell in the sea, the proper number for each boat could not, with propriety, be rated at more than twelve persons.

When the tide-bell rung the boats were hoisted out, and two active seamen were employed to keep them from receiving damage alongside. The floating light

1807 being very buoyant, was so quick in her motions that when those who were about to step from her gunwale into a boat, placed themselves upon a cleat or step on the ship's side, with the man or rail ropes in their hands, they had often to wait for some time till a favourable opportunity occurred for stepping into the boat. While in this situation, with the vessel rolling from side to side, watching the proper time for letting go the man-ropes, it required the greatest dexterity and presence of mind to leap into the boats. One who was rather awkward would often wait a considerable period in this position: at one time his side of the ship would be so depressed that he would touch the boat to which he belonged, while the next sea would elevate him so much that he would see his comrades in the boat on the opposite side of the ship, his friends in the one boat calling to him to 'Jump,' while those in the boat on the other side, as he came again and again into their view, would jocosely say, 'Are you there yet? You seem to enjoy a swing.' In this situation it was common to see a person upon each side of the ship for a length of time, waiting to quit his hold.

On leaving the rock to-day a trial of seamanship was proposed amongst the rowers, for by this time the artificers had become tolerably expert in this exercise. By inadvertency some of the oars provided had been made of fir instead of ash, and although a considerable stock had been laid in, the workmen, being at first awkward in the art, were constantly breaking their oars; indeed it was no uncommon thing to see the broken blades of a pair of oars floating astern, in the course of a passage from the rock to the vessel. The men, upon the whole, had but little work to perform in the course of a day; for though they exerted them-

1807

selves extremely hard while on the rock, yet, in the early state of the operations, this could not be continued for more than three or four hours at a time, and as their rations were large—consisting of one pound and a half of beef, one pound of ship biscuit, eight ounces oatmeal, two ounces barley, two ounces butter, three quarts of small beer, with vegetables and salt—they got into excellent spirits when free of sea-sickness. The rowing of the boats against each other became a favourite amusement, which was rather a fortunate circumstance, as it must have been attended with much inconvenience had it been found necessary to employ a sufficient number of sailors for this purpose. The writer, therefore, encouraged the spirit of emulation, and the speed of their respective boats became a favourite topic. Premiums for boat-races were instituted, which were contended for with great eagerness, and the respective crews kept their stations in the boats with as much precision as they kept their beds on board of the ship. With these and other pastimes, when the weather was favourable, the time passed away among the inmates of the forecastle and waist of the ship. The writer looks back with interest upon the hours of solitude which he spent in this lonely ship with his small library.

This being the first Saturday that the artificers were afloat, all hands were served with a glass of rum and water at night, to drink the sailors' favourite toast of 'Wives and Sweethearts.' It was customary, upon these occasions, for the seamen and artificers to collect in the galley, when the musical instruments were put in requisition: for, according to invariable practice, every man must play a tune, sing a song, or tell a story.

Sunday, 23rd Aug.

Having, on the previous evening, arranged matters

1807 with the landing-master as to the business of the day, the signal was rung for all hands at half-past seven this morning. In the early state of the spring-tides the artificers went to the rock before breakfast, but as the tides fell later in the day, it became necessary to take this meal before leaving the ship. At eight o'clock all hands were assembled on the quarter-deck for prayers, a solemnity which was gone through in as orderly a manner as circumstances would admit. When the weather permitted, the flags of the ship were hung up as an awning or screen, forming the quarter-deck into a distinct compartment; the pendant was also hoisted at the mainmast, and a large ensign flag was displayed over the stern; and lastly, the ship's companion, or top of the staircase, was covered with the *flag proper* of the Lighthouse Service, on which the Bible was laid. A particular toll of the bell called all hands to the quarter-deck, when the writer read a chapter of the Bible, and, the whole ship's company being uncovered, he also read the impressive prayer composed by the Reverend Dr. Brunton, one of the ministers of Edinburgh.

Upon concluding this service, which was attended with becoming reverence and attention, all on board retired to their respective berths to breakfast, and, at half-past nine, the bell again rung for the artificers to take their stations in their respective boats. Some demur having been evinced on board about the propriety of working on Sunday, which had hitherto been touched upon as delicately as possible, all hands being called aft, the writer, from the quarter-deck, stated generally the nature of the service, expressing his hopes that every man would feel himself called upon to consider the erection of a lighthouse on the Bell Rock, in every point of view, as a work of necessity and mercy. He knew that scruples had existed with some, and these had, indeed, been fairly and candidly urged before leaving the shore; but it was expected that, after having seen the critical nature of the rock, and the necessity of the measure, every man would now be satisfied of the propriety of embracing all opportunities of landing on the rock when the state of the weather would permit. The writer further took them to witness that it did not proceed from want of respect for the appointments and established forms of religion that he had himself adopted the resolution of attending the Bell Rock works on the Sunday; but, as he hoped, from a conviction that it was his bounden duty, on the strictest principles of morality. At the same time it was intimated that, if any were of a different opinion, they should be perfectly at liberty to hold their sentiments without the imputation of contumacy or disobedience; the only difference would be in regard to the pay.

1807

Upon stating this much, he stepped into his boat, requesting all who were so disposed to follow him. The sailors, from their habits, found no scruple on this subject, and all of the artificers, though a little tardy, also embarked, excepting four of the masons, who, from the beginning, mentioned that they would decline working on Sundays. It may here be noticed that throughout the whole of the operations it was observable that the men wrought, if possible, with more keenness upon the Sundays than at other times, from an impression that they were engaged in a work of imperious necessity, which required every possible exertion. On returning to the floating light, after finishing the tide's work, the boats were received by the part of the ship's crew left on board with the usual attention of handing ropes to the boats and

1807 helping the artificers on board; but the four masons who had absented themselves from the work did not appear upon deck.

Monday, 24th Aug. The boats left the floating light at a quarter-past nine o'clock this morning, and the work began at three-quarters past nine; but as the neap-tides were approaching the working time at the rock became gradually shorter, and it was now with difficulty that two and a half hours' work could be got. But so keenly had the workmen entered into the spirit of the beacon-house operations, that they continued to bore the holes in the rock till some of them were knee-deep in water.

The operations at this time were entirely directed to the erection of the beacon, in which every man felt an equal interest, as at this critical period the slightest casualty to any of the boats at the rock might have been fatal to himself individually, while it was perhaps peculiar to the writer more immediately to feel for the safety of the whole. Each log or upright beam of the beacon was to be fixed to the rock by two strong and massive bats or stanchions of iron. These bats, for the fixture of the principal and diagonal beams and bracing chains, required fifty-four holes, each measuring two inches in diameter and eighteen inches in depth. There had already been so considerable a progress made in boring and excavating the holes that the writer's hopes of getting the beacon erected this year began to be more and more confirmed, although it was now advancing towards what was considered the latter end of the proper working season at the Bell Rock. The foreman joiner, Mr. Francis Watt, was accordingly appointed to attend at the rock to-day, when the necessary levels were taken for the step or seat of each particular beam of the beacon, that they might

1807

be cut to their respective lengths, to suit the inequalities of the rock; several of the stanchions were also tried into their places, and other necessary observations made, to prevent mistakes on the application of the apparatus, and to facilitate the operations when the beams came to be set up, which would require to be done in the course of a single tide.

Tuesday, 25th Aug.

We had now experienced an almost unvaried tract of light airs of easterly wind, with clear weather in the fore-part of the day and fog in the evenings. To-day, however, it sensibly changed; when the wind came to the south-west, and blew a fresh breeze. At nine a.m. the bell rung, and the boats were hoisted out, and though the artificers were now pretty well accustomed to tripping up and down the sides of the floating light, yet it required more seamanship this morning than usual. It therefore afforded some merriment to those who had got fairly seated in their respective boats to see the difficulties which attended their companions, and the hesitating manner in which they quitted hold of the man-ropes in leaving the ship. The passage to the rock was tedious, and the boats did not reach it till half-past ten.

It being now the period of neap-tides, the water only partially left the rock, and some of the men who were boring on the lower ledges of the site of the beacon stood knee-deep in water. The situation of the smith to-day was particularly disagreeable, but his services were at all times indispensable. As the tide did not leave the site of the forge, he stood in the water, and as there was some roughness on the surface it was with considerable difficulty that, with the assistance of the sailors, he was enabled to preserve alive his fire; and, while his feet were immersed in water, his face was not only scorched but continually

1807 exposed to volumes of smoke, accompanied with sparks from the fire, which were occasionally set up owing to the strength and direction of the wind.

Wednesday, 26th Aug. The wind had shifted this morning to N.N.W., with rain, and was blowing what sailors call a fresh breeze. To speak, perhaps, somewhat more intelligibly to the general reader, the wind was such that a fishing-boat could just carry full sail. But as it was of importance, specially in the outset of the business, to keep up the spirit of enterprise for landing on all practicable occasions, the writer, after consulting with the landing-master, ordered the bell to be rung for embarking, and at half-past eleven the boats reached the rock, and left it again at a quarter-past twelve, without, however, being able to do much work, as the smith could not be set to work from the smallness of the ebb and the strong breach of sea, which lashed with great force among the bars of the forge.

Just as we were about to leave the rock the wind shifted to the S.W., and, from a fresh gale, it became what seamen term a hard gale, or such as would have required the fisherman to take in two or three reefs in his sail. It is a curious fact that the respective tides of ebb and flood are apparent upon the shore about an hour and a half sooner than at the distance of three or four miles in the offing. But what seems chiefly interesting here is that the tides around this small sunken rock should follow exactly the same laws as on the extensive shores of the mainland. When the boats left the Bell Rock to-day it was overflowed by the flood-tide, but the floating light did not swing round to the flood-tide for more than an hour afterwards. Under this disadvantage the boats had to struggle with the ebb-tide and a hard gale of wind, so that it was with the greatest difficulty that they reached the

1807

floating light. Had this gale happened in spring-tides when the current was strong we must have been driven to sea in a very helpless condition.

The boat which the writer steered was considerably behind the other, one of the masons having unluckily broken his oar. Our prospect of getting on board, of course, became doubtful, and our situation was rather perilous, as the boat shipped so much sea that it occupied two of the artificers to bale and clear her of water. When the oar gave way we were about half a mile from the ship, but, being fortunately to windward, we got into the wake of the floating light, at about 250 fathoms astern, just as the landing-master's boat reached the vessel. He immediately streamed or floated a life-buoy astern, with a line which was always in readiness, and by means of this useful implement the boat was towed alongside of the floating light, where, from her rolling motion, it required no small management to get safely on board, as the men were much worn out with their exertions in pulling from the rock. On the present occasion the crews of both boats were completely drenched with spray, and those who sat upon the bottom of the boats to bale them were sometimes pretty deep in the water before it could be cleared out. After getting on board, all hands were allowed an extra dram, and, having shifted and got a warm and comfortable dinner, the affair, it is believed, was little more thought of.

Thursday, 27th Aug.

The tides were now in that state which sailors term the dead of the neap, and it was not expected that any part of the rock would be seen above water to-day; at any rate, it was obvious, from the experience of yesterday, that no work could be done upon it, and therefore the artificers were not required to land. The wind was at west, with light breezes, and fine clear weather;

1807 and as it was an object with the writer to know the actual state of the Bell Rock at neap-tides, he got one of the boats manned, and, being accompanied by the landing-master, went to it at a quarter-past twelve. The parts of the rock that appeared above water being very trifling, were covered by every wave, so that no landing was made. Upon trying the depth of water with a boathook, particularly on the sites of the lighthouse and beacon, on the former, at low water, the depth was found to be three feet, and on the central parts of the latter it was ascertained to be two feet eight inches. Having made these remarks, the boat returned to the ship at two p.m., and the weather being good, the artificers were found amusing themselves with fishing. The *Smeaton* came from Arbroath this afternoon, and made fast to her moorings, having brought letters and newspapers, with parcels of clean linen, etc., for the workmen, who were also made happy by the arrival of three of their comrades from the workyard ashore. From these men they not only received all the news of the workyard, but seemed themselves to enjoy great pleasure in communicating whatever they considered to be interesting with regard to the rock. Some also got letters from their friends at a distance, the postage of which for the men afloat was always free, so that they corresponded the more readily.

The site of the building having already been carefully traced out with the pick-axe, the artificers this day commenced the excavation of the rock for the foundation or first course of the lighthouse. Four men only were employed at this work, while twelve continued at the site of the beacon-house, at which every possible opportunity was embraced, till this essential part of the operations should be completed.

1807 Wednesday, 2nd Sept.

The floating light's bell rung this morning at half-past four o'clock, as a signal for the boats to be got ready, and the landing took place at half-past five. In passing the *Smeaton* at her moorings near the rock, her boat followed with eight additional artificers who had come from Arbroath with her at last trip, but there being no room for them in the floating light's boats, they had continued on board. The weather did not look very promising in the morning, the wind blowing pretty fresh from W.S.W.: and had it not been that the writer calculated upon having a vessel so much at command, in all probability he would not have ventured to land. The *Smeaton* rode at what sailors call a *salvagee*, with a cross-head made fast to the floating buoy. This kind of attachment was found to be more convenient than the mode of passing the hawser through the ring of the buoy when the vessel was to be made fast. She had then only to be steered very close to the buoy, when the salvagee was laid hold of with a boat-hook, and the *bite* of the hawser thrown over the cross-head. But the salvagee, by this method, was always left at the buoy, and was, of course, more liable to chafe and wear than a hawser passed through the ring, which could be wattled with canvas, and shifted at pleasure. The salvagee and cross method is, however, much practised; but the experience of this morning showed it to be very unsuitable for vessels riding in an exposed situation for any length of time.

Soon after the artificers landed they commenced work; but the wind coming to blow hard, the *Smeaton's* boat and crew, who had brought their complement of eight men to the rock, went off to examine her riding ropes, and see that they were in proper order. The boat had no sooner reached the vessel

1807 than she went adrift, carrying the boat along with her. By the time that she was got round to make a tack towards the rock, she had drifted at least three miles to leeward, with the praam-boat astern; and, having both the wind and a tide against her, the writer perceived, with no little anxiety, that she could not possibly return to the rock till long after its being overflowed; for, owing to the anomaly of the tides formerly noticed, the Bell Rock is completely under water when the ebb abates to the offing.

In this perilous predicament, indeed, he found himself placed between hope and despair—but certainly the latter was by much the most predominant feeling of his mind—situate upon a sunken rock in the middle of the ocean, which, in the progress of the flood-tide, was to be laid under water to the depth of at least twelve feet in a stormy sea. There were this morning thirty-two persons in all upon the rock, with only two boats, whose complement, even in good weather, did not exceed twenty-four sitters; but to row to the floating light with so much wind, and in so heavy a sea, a complement of eight men for each boat was as much as could, with propriety, be attempted, so that, in this way, about one-half of our number was unprovided for. Under these circumstances, had the writer ventured to despatch one of the boats in expectation of either working the *Smeaton* sooner up towards the rock, or in hopes of getting her boat brought to our assistance, this must have given an immediate alarm to the artificers, each of whom would have insisted upon taking to his own boat, and leaving the eight artificers belonging to the *Smeaton* to their chance. Of course a scuffle might have ensued, and it is hard to say, in the ardour of men contending for life, where it might have ended. It has even been hinted to the writer that a party of the *pickmen* were determined to keep exclusively to their own boat against all hazards.

The unfortunate circumstance of the *Smeaton* and her boat having drifted was, for a considerable time, only known to the writer and to the landing-master, who removed to the farther point of the rock, where he kept his eye steadily upon the progress of the vessel. While the artificers were at work, chiefly in sitting or kneeling postures, excavating the rock, or boring with the jumpers, and while their numerous hammers, with the sound of the smith's anvil, continued, the situation of things did not appear so awful. In this state of suspense, with almost certain destruction at hand, the water began to rise upon those who were at work on the lower parts of the sites of the beacon and lighthouse. From the run of sea upon the rock, the forge fire was also sooner extinguished this morning than usual, and the volumes of smoke having ceased, objects in every direction became visible from all parts of the rock. After having had about three hours' work, the men began, pretty generally, to make towards their respective boats for their jackets and stockings, when, to their astonishment, instead of three, they found only two boats, the third being adrift with the *Smeaton*. Not a word was uttered by any one, but all appeared to be silently calculating their numbers, and looking to each other with evident marks of perplexity depicted in their countenances. The landing-master, conceiving that blame might be attached to him for allowing the boat to leave the rock, still kept at a distance. At this critical moment the author was standing upon an elevated part of Smith's Ledge, where he endeavoured to mark the progress of the *Smeaton*, not a little surprised that her

1807 crew did not cut the praam adrift, which greatly retarded her way, and amazed that some effort was not making to bring at least the boat, and attempt our relief. The workmen looked steadfastly upon the writer, and turned occasionally towards the vessel, still far to leeward.[1] All this passed in the most perfect silence, and the melancholy solemnity of the group made an impression never to be effaced from his mind.

The writer had all along been considering of various schemes—providing the men could be kept under command—which might be put in practice for the general safety, in hopes that the *Smeaton* might be able to pick up the boats to leeward, when they were obliged to leave the rock. He was, accordingly, about to address the artificers on the perilous nature of their circumstances, and to propose that all hands should unstrip their upper clothing when the higher parts of the rock were laid under water; that the seamen should remove every unnecessary weight and encumbrance from the boats; that a specified number of men should go into each boat, and that the remainder should hang by the gunwales, while the boats were to be rowed gently towards the *Smeaton,* as the course to the *Pharos,* or floating light, lay rather to windward of the rock. But when he attempted to speak his mouth was so parched that his tongue refused utterance, and he now learned by experience that the saliva is as necessary as the tongue itself for speech. He turned to one of the pools on the rock and lapped a little water, which produced immediate relief. But what was his happiness, when on rising from this unpleasant beverage, some one called out,

[1] 'Nothing was said, but I was *looked out of countenance,*' he says in a letter.

'A boat! a boat!' and, on looking around, at no great distance, a large boat was seen through the haze making towards the rock. This at once enlivened and rejoiced every heart. The timeous visitor proved to be James Spink, the Bell Rock pilot, who had come express from Arbroath with letters. Spink had for some time seen the *Smeaton,* and had even supposed, from the state of the weather, that all hands were on board of her till he approached more nearly and observed people upon the rock; but not supposing that the assistance of his boat was necessary to carry the artificers off the rock, he anchored on the lee-side and began to fish, waiting, as usual, till the letters were sent for, as the pilot-boat was too large and unwieldy for approaching the rock when there was any roughness or run of the sea at the entrance of the landing creeks.

Upon this fortunate change of circumstances, sixteen of the artificers were sent, at two trips, in one of the boats, with instructions for Spink to proceed with them to the floating light. This being accomplished, the remaining sixteen followed in the two boats belonging to the service of the rock. Every one felt the most perfect happiness at leaving the Bell Rock this morning, though a very hard and even dangerous passage to the floating light still awaited us, as the wind by this time had increased to a pretty hard gale, accompanied with a considerable swell of sea. Every one was as completely drenched in water as if he had been dragged astern of the boats. The writer, in particular, being at the helm, found, on getting on board, that his face and ears were completely coated with a thin film of salt from the sea spray, which broke constantly over the bows of the boat. After much baling of water and severe work at the oars, the

1807 three boats reached the floating light, where some new difficulties occurred in getting on board in safety, owing partly to the exhausted state of the men, and partly to the violent rolling of the vessel.

As the tide flowed, it was expected that the *Smeaton* would have got to windward; but, seeing that all was safe, after tacking for several hours and making little progress, she bore away for Arbroath, with the praam-boat. As there was now too much wind for the pilot-boat to return to Arbroath, she was made fast astern of the floating light, and the crew remained on board till next day, when the weather moderated. There can be very little doubt that the appearance of James Spink with his boat on this critical occasion was the means of preventing the loss of lives at the rock this morning. When these circumstances, some years afterwards, came to the knowledge of the Board, a small pension was ordered to our faithful pilot, then in his seventieth year; and he still continues to wear the uniform clothes and badge of the Lighthouse service. Spink is a remarkably strong man, whose *tout ensemble* is highly characteristic of a North-country fisherman. He usually dresses in a *pé-jacket*, cut after a particular fashion, and wears a large, flat, blue bonnet. A striking likeness of Spink in his pilot-dress, with the badge or insignia on his left arm which is characteristic of the boatmen in the service of the Northern Lights, has been taken by Howe, and is in the writer's possession.

Thursday, 3rd Sept.

The bell rung this morning at five o'clock, but the writer must acknowledge, from the circumstances of yesterday, that its sound was extremely unwelcome. This appears also to have been the feelings of the artificers, for when they came to be mustered, out of twenty-six, only eight, besides the foreman and sea-

1807

men, appeared upon deck to accompany the writer to the rock. Such are the baneful effects of anything like misfortune or accident connected with a work of this description. The use of argument to persuade the men to embark in cases of this kind would have been out of place, as it is not only discomfort, or even the risk of the loss of a limb, but life itself that becomes the question. The boats, notwithstanding the thinness of our ranks, left the vessel at half-past five. The rough weather of yesterday having proved but a summer's gale, the wind came to-day in gentle breezes; yet, the atmosphere being cloudy, it had not a very favourable appearance. The boats reached the rock at six a.m., and the eight artificers who landed were employed in clearing out the bat-holes for the beacon-house, and had a very prosperous tide of four hours' work, being the longest yet experienced by half an hour.

The boats left the rock again at ten o'clock, and the weather having cleared up as we drew near the vessel, the eighteen artificers who had remained on board were observed upon deck, but as the boats approached they sought their way below, being quite ashamed of their conduct. This was the only instance of refusal to go to the rock which occurred during the whole progress of the work, excepting that of the four men who declined working upon Sunday, a case which the writer did not conceive to be at all analogous to the present. It may here be mentioned, much to the credit of these four men, that they stood foremost in embarking for the rock this morning.

Saturday, 5th Sept.

It was fortunate that a landing was not attempted this evening, for at eight o'clock the wind shifted to E.S.E., and at ten it had become a hard gale, when fifty fathoms of the floating light's hempen cable were

1807 veered out. The gale still increasing, the ship rolled and laboured excessively, and at midnight eighty fathoms of cable were veered out; while the sea continued to strike the vessel with a degree of force which had not before been experienced.

Sunday, 6th Sept.

During the last night there was little rest on board of the *Pharos*, and daylight, though anxiously wished for, brought no relief, as the gale continued with unabated violence. The sea struck so hard upon the vessel's bows that it rose in great quantities, or in 'green seas,' as the sailors termed it, which were carried by the wind as far aft as the quarter-deck, and not infrequently over the stern of the ship altogether. It fell occasionally so heavily on the skylight of the writer's cabin, though so far aft as to be within five feet of the helm, that the glass was broken to pieces before the dead-light could be got into its place, so that the water poured down in great quantities. In shutting out the water, the admission of light was prevented, and in the morning all continued in the most comfortless state of darkness. About ten o'clock a.m. the wind shifted to N.E., and blew, if possible, harder than before, and it was accompanied by a much heavier swell of sea. In the course of the gale, the part of the cable in the hause-hole had been so often shifted that nearly the whole length of one of her hempen cables, of 120 fathoms, had been veered out, besides the chain-moorings. The cable, for its preservation, was also carefully served or wattled with pieces of canvas round the windlass, and with leather well greased in the hause-hole. In this state things remained during the whole day, every sea which struck the vessel—and the seas followed each other in close succession—causing her to shake, and all on board occasionally to tremble. At each of these strokes of the sea the rolling and pitching of the vessel ceased for a time, and her motion was felt as if she had either broke adrift before the wind or were in the act of sinking; but, when another sea came, she ranged up against it with great force, and this became the regular intimation of our being still riding at anchor.

About eleven o'clock, the writer with some difficulty got out of bed, but, in attempting to dress, he was thrown twice upon the floor at the opposite end of the cabin. In an undressed state he made shift to get about half-way up the companion-stairs, with an intention to observe the state of the sea and of the ship upon deck; but he no sooner looked over the companion than a heavy sea struck the vessel, which fell on the quarter-deck, and rushed downstairs in the officers' cabin in so considerable a quantity that it was found necessary to lift one of the scuttles in the floor, to let the water into the limbers of the ship, as it dashed from side to side in such a manner as to run into the lower tier of beds. Having been foiled in this attempt, and being completely wetted, he again got below and went to bed. In this state of the weather the seamen had to move about the necessary or indispensable duties of the ship with the most cautious use both of hands and feet, while it required all the art of the landsman to keep within the precincts of his bed. The writer even found himself so much tossed about that it became necessary, in some measure, to shut himself in bed, in order to avoid being thrown upon the floor. Indeed, such was the motion of the ship that it seemed wholly impracticable to remain in any other than a lying posture. On deck the most stormy aspect presented itself, while below all was wet and comfortless.

About two o'clock p.m. a great alarm was given

1807 throughout the ship from the effects of a very heavy sea which struck her, and almost filled the waist, pouring down into the berths below, through every chink and crevice of the hatches and skylights. From the motion of the vessel being thus suddenly deadened or checked, and from the flowing in of the water above, it is believed there was not an individual on board who did not think, at the moment, that the vessel had foundered, and was in the act of sinking. The writer could withstand this no longer, and as soon as she again began to range to the sea he determined to make another effort to get upon deck. In the first instance, however, he groped his way in darkness from his own cabin through the berths of the officers, where all was quietness. He next entered the galley and other compartments occupied by the artificers. Here also all was shut up in darkness, the fire having been drowned out in the early part of the gale. Several of the artificers were employed in prayer, repeating psalms and other devotional exercises in a full tone of voice; others protesting that, if they should fortunately get once more on shore, no one should ever see them afloat again. With the assistance of the landing-master, the writer made his way, holding on step by step, among the numerous impediments which lay in the way. Such was the creaking noise of the bulkheads or partitions, the dashing of the water, and the whistling noise of the winds, that it was hardly possible to break in upon such a confusion of sounds. In one or two instances, anxious and repeated inquiries were made by the artificers as to the state of things upon deck, to which the captain made the usual answer, that it could not blow long in this way, and that we must soon have better weather. The next berth in succession, moving forward in the ship, was that allotted for the seamen. Here the scene was considerably different. Having reached the middle of this darksome berth without its inmates being aware of any intrusion, the writer had the consolation of remarking that, although they talked of bad weather and the cross accidents of the sea, yet the conversation was carried on in that sort of tone and manner which bespoke an ease and composure of mind highly creditable to them and pleasing to him. The writer immediately accosted the seamen about the state of the ship. To these inquiries they replied that the vessel being light, and having but little hold of the water, no top-rigging, with excellent ground-tackle, and everything being fresh and new, they felt perfect confidence in their situation.

It being impossible to open any of the hatches in the fore part of the ship in communicating with the deck, the watch was changed by passing through the several berths to the companion-stair leading to the quarter-deck. The writer, therefore, made the best of his way aft, and, on a second attempt to look out, he succeeded, and saw indeed an astonishing sight. The sea or waves appeared to be ten or fifteen feet in height of unbroken water, and every approaching billow seemed as if it would overwhelm our vessel, but she continued to rise upon the waves and to fall between the seas in a very wonderful manner. It seemed to be only those seas which caught her in the act of rising which struck her with so much violence and threw such quantities of water aft. On deck there was only one solitary individual looking out. to give the alarm in the event of the ship breaking from her moorings. The seaman on watch continued only two hours; he who kept watch at this time was a tall, slender man of a black complexion; he had no greatcoat nor over-all of any

1807 kind, but was simply dressed in his ordinary jacket and trousers; his hat was tied under his chin with a napkin, and he stood aft the foremast, to which he had lashed himself with a gasket or small rope round his waist, to prevent his falling upon deck or being washed overboard. When the writer looked up, he appeared to smile, which afforded a further symptom of the confidence of the crew in their ship. This person on watch was as completely wetted as if he had been drawn through the sea, which was given as a reason for his not putting on a greatcoat, that he might wet as few of his clothes as possible, and have a dry shift when he went below. Upon deck everything that was movable was out of sight, having either been stowed below, previous to the gale, or been washed overboard. Some trifling parts of the quarter boards were damaged by the breach of the sea; and one of the boats upon deck was about one-third full of water, the oyle-hole or drain having been accidently stopped up, and part of her gunwale had received considerable injury. These observations were hastily made, and not without occasionally shutting the companion, to avoid being wetted by the successive seas which broke over the bows and fell upon different parts of the deck according to the impetus with which the waves struck the vessel. By this time it was about three o'clock in the afternoon, and the gale, which had now continued with unabated force for twenty-seven hours, had not the least appearance of going off.

In the dismal prospect of undergoing another night like the last, and being in imminent hazard of parting from our cable, the writer thought it necessary to advise with the master and officers of the ship as to the probable event of the vessel's drifting from her moorings. They severally gave it as their opinion that

1807

we had now every chance of riding out the gale, which, in all probability, could not continue with the same fury many hours longer; and that even if she should part from her anchor, the storm-sails had been laid to hand, and could be bent in a very short time. They further stated that from the direction of the wind being N.E., she would sail up the Firth of Forth to Leith Roads. But if this should appear doubtful, after passing the Island and Light of May, it might be advisable at once to steer for Tyningham Sands, on the western side of Dunbar, and there run the vessel ashore. If this should happen at the time of high-water, or during the ebbing of the tide, they were of opinion, from the flatness and strength of the floating light, that no danger would attend her taking the ground, even with a very heavy sea. The writer, seeing the confidence which these gentlemen possessed with regard to the situation of things, found himself as much relieved with this conversation as he had previously been with the seeming indifference of the forecastle men, and the smile of the watch upon deck, though literally lashed to the foremast. From this time he felt himself almost perfectly at ease; at any rate, he was entirely resigned to the ultimate result.

About six o'clock in the evening the ship s company was heard moving upon deck, which on the present occasion was rather the cause of alarm. The writer accordingly rang his bell to know what was the matter, when he was informed by the steward that the weather looked considerably better, and that the men upon deck were endeavouring to ship the smoke-funnel of the galley that the people might get some meat. This was a more favourable account than had been anticipated. During the last twenty-one hours he himself had not only had nothing to eat, but he had

1807 almost never passed a thought on the subject. Upon the mention of a change of weather, he sent the steward to learn how the artificers felt, and on his return he stated that they now seemed to be all very happy, since the cook had begun to light the galley-fire and make preparations for the suet-pudding of Sunday, which was the only dish to be attempted for the mess, from the ease with which it could both be cooked and served up.

The principal change felt upon the ship as the wind abated was her increased rolling motion, but the pitching was much diminished, and now hardly any sea came farther aft than the foremast: but she rolled so extremely hard as frequently to dip and take in water over the gunwales and rails in the waist. By nine o'clock all hands had been refreshed by the exertions of the cook and steward, and were happy in the prospect of the worst of the gale being over. The usual complement of men was also now set on watch, and more quietness was experienced throughout the ship. Although the previous night had been a very restless one, it had not the effect of inducing repose in the writer's berth on the succeeding night; for having been so much tossed about in bed during the last thirty hours, he found no easy spot to turn to, and his body was all sore to the touch, which ill accorded with the unyielding materials with which his bed-place was surrounded.

Monday, 7th Sept. This morning, about eight o'clock, the writer was agreeably surprised to see the scuttle of his cabin sky-light removed, and the bright rays of the sun admitted. Although the ship continued to roll excessively, and the sea was still running very high, yet the ordinary business on board seemed to be going forward on deck. It was impossible to steady a telescope, so as to look

1807

minutely at the progress of the waves and trace their breach upon the Bell Rock; but the height to which the cross-running waves rose in sprays when they met each other was truly grand, and the continued roar and noise of the sea was very perceptible to the ear. To estimate the height of the sprays at forty or fifty feet would surely be within the mark. Those of the workmen who were not much afflicted with sea-sickness, came upon deck, and the wetness below being dried up, the cabins were again brought into a habitable state. Every one seemed to meet as if after a long absence, congratulating his neighbour upon the return of good weather. Little could be said as to the comfort of the vessel, but after riding out such a gale, no one felt the least doubt or hesitation as to the safety and good condition of her moorings. The master and mate were extremely anxious, however, to heave in the hempen cable, and see the state of the clinch or iron ring of the chain-cable. But the vessel rolled at such a rate that the seamen could not possibly keep their feet at the windlass nor work the handspikes, though it had been several times attempted since the gale took off.

About twelve noon, however, the vessel's motion was observed to be considerably less, and the sailors were enabled to walk upon deck with some degree of freedom. But, to the astonishment of every one, it was soon discovered that the floating light was adrift! The windlass was instantly manned, and the men soon gave out that there was no strain upon the cable. The mizzen sail, which was bent for the occasional purpose of making the vessel ride more easily to the tide, was immediately set, and the other sails were also hoisted in a short time, when, in no small consternation, we bore away about one mile to

1807 the south-westward of the former station, and there let go the best bower anchor and cable in twenty fathoms water, to ride until the swell of the sea should fall, when it might be practicable to grapple for the moorings, and find a better anchorage for the ship.

Tuesday, 15th Sept. This morning, at five a.m., the bell rung as a signal for landing upon the rock, a sound which, after a lapse of ten days, it is believed was welcomed by every one on board. There being a heavy breach of sea at the eastern creek, we landed, though not without difficulty, on the western side, every one seeming more eager than another to get upon the rock; and never did hungry men sit down to a hearty meal with more appetite than the artificers began to pick the dulse from the rocks. This marine plant had the effect of reviving the sickly, and seemed to be no less relished by those who were more hardy.

While the water was ebbing, and the men were roaming in quest of their favourite morsel, the writer was examining the effects of the storm upon the forge and loose apparatus left upon the rock. Six large blocks of granite which had been landed, by way of experiment, on the 1st instant, were now removed from their places and, by the force of the sea, thrown over a rising ledge into a hole at the distance of twelve or fifteen paces from the place on which they had been landed. This was a pretty good evidence both of the violence of the storm and the agitation of the sea upon the rock. The safety of the smith's forge was always an object of essential regard. The ash-pan of the hearth or fireplace, with its weighty cast-iron back, had been washed from their places of supposed security; the chains of attachment had been broken, and these ponderous articles were found at a very considerable distance in a hole on the western side of the rock; while the tools and picks of the Aberdeen masons were scattered about in every direction. It is, however, remarkable that not a single article was ultimately lost.

This being the night on which the floating light was advertised to be lighted, it was accordingly exhibited, to the great joy of every one.

Wednesday, 16th Sept.

The writer was made happy to-day by the return of the Lighthouse yacht from a voyage to the Northern Lighthouses. Having immediately removed on board of this fine vessel of eighty-one tons register, the artificers gladly followed; for, though they found themselves more pinched for accommodation on board of the yacht, and still more so in the *Smeaton*, yet they greatly preferred either of these to the *Pharos*, or floating light, on account of her rolling motion, though in all respects fitted up for their conveniency.

The writer called them to the quarter-deck and informed them that, having been one month afloat, in terms of their agreement they were now at liberty to return to the workyard at Arbroath if they preferred this to continuing at the Bell Rock. But they replied that, in the prospect of soon getting the beacon erected upon the rock, and having made a change from the floating light, they were now perfectly reconciled to their situation, and would remain afloat till the end of the working season.

Thursday, 17th Sept.

The wind was at N.E. this morning, and though they were only light airs, yet there was a pretty heavy swell coming ashore upon the rock. The boats landed at half-past seven o'clock a.m., at the creek on the southern side of the rock, marked Port Hamilton. But as one of the boats was in the act of entering this creek, the seaman at the bow-oar, who had just entered the service, having inadvertently expressed some fear from

1807 a heavy sea which came rolling towards the boat, and one of the artificers having at the same time looked round and missed a stroke with his oar, such a preponderance was thus given to the rowers upon the opposite side that when the wave struck the boat it threw her upon a ledge of shelving rocks, where the water left her, and she having *kanted* to seaward, the next wave completely filled her with water. After making considerable efforts the boat was again got afloat in the proper track of the creek, so that we landed without any other accident than a complete ducking. There being no possibility of getting a shift of clothes, the artificers began with all speed to work, so as to bring themselves into heat, while the writer and his assistants kept as much as possible in motion. Having remained more than an hour upon the rock, the boats left it at half-past nine; and, after getting on board, the writer recommended to the artificers, as the best mode of getting into a state of comfort, to strip off their wet clothes and go to bed for an hour or two. No further inconveniency was felt, and no one seemed to complain of the affection called 'catching cold.'

Friday, 18th Sept. An important occurrence connected with the operations of this season was the arrival of the *Smeaton* at four p.m., having in tow the six principal beams of the beacon-house, together with all the stanchions and other work on board for fixing it on the rock. The mooring of the floating light was a great point gained, but in the erection of the beacon at this late period of the season new difficulties presented themselves. The success of such an undertaking at any season was precarious, because a single day of bad weather occurring before the necessary fixtures could be made might sweep the whole apparatus from the rock. Notwith-

1807

standing these difficulties, the writer had determined to make the trial, although he could almost have wished, upon looking at the state of the clouds and the direction of the wind, that the apparatus for the beacon had been still in the workyard.

Saturday, 19th Sept.

The main beams of the beacon were made up in two separate rafts, fixed with bars and bolts of iron. One of these rafts, not being immediately wanted, was left astern of the floating light, and the other was kept in tow by the *Smeaton*, at the buoy nearest to the rock. The Lighthouse yacht rode at another buoy with all hands on board that could possibly be spared out of the floating light. The party of artificers and seamen which landed on the rock counted altogether forty in number. At half-past eight o'clock a derrick, or mast of thirty feet in height, was erected and properly supported with guy-ropes, for suspending the block for raising the first principal beam of the beacon; and a winch machine was also bolted down to the rock for working the purchase-tackle.

Upon raising the derrick, all hands on the rock spontaneously gave three hearty cheers, as a favourable omen of our future exertions in pointing out more permanently the position of the rock. Even to this single spar of timber, could it be preserved, a drowning man might lay hold. When the *Smeaton* drifted on the 2nd of this month such a spar would have been sufficient to save us till she could have come to our relief.

Sunday, 20th Sept.

The wind this morning was variable, but the weather continued extremely favourable for the operations throughout the whole day. At six a.m. the boats were in motion, and the raft, consisting of four of the six principal beams of the beacon-house,

1807 each measuring about sixteen inches square, and fifty feet in length, was towed to the rock, where it was anchored, that it might *ground* upon it as the water ebbed. The sailors and artificers, including all hands, to-day counted no fewer than fifty-two, being perhaps the greatest number of persons ever collected upon the Bell Rock. It was early in the tide when the boats reached the rock, and the men worked a considerable time up to their middle in water, every one being more eager than his neighbour to be useful. Even the four artificers who had hitherto declined working on Sunday were to-day most zealous in their exertions. They had indeed become so convinced of the precarious nature and necessity of the work that they never afterwards absented themselves from the rock on Sunday when a landing was practicable.

Having made fast a piece of very good new line, at about two-thirds from the lower end of one of the beams, the purchase-tackle of the derrick was hooked into the turns of the line, and it was speedily raised by the number of men on the rock and the power of the winch tackle. When this log was lifted to a sufficient height, its foot, or lower end, was *stepped* into the spot which had been previously prepared for it. Two of the great iron stanchions were then set in their respective holes on each side of the beam, when a rope was passed round them and the beam, to prevent it from slipping till it could be more permanently fixed. The derrick, or upright spar used for carrying the tackle to raise the first beam, was placed in such a position as to become useful for supporting the upper end of it, which now became, in its turn, the prop of the tackle for raising the second beam. The whole difficulty of this operation was in the raising and propping of the first beam, which became a convenient derrick for raising the second, these again a pair of shears for lifting the third, and the shears a triangle for raising the fourth. Having thus got four of the six principal beams set on end, it required a considerable degree of trouble to get their upper ends to fit. Here they formed the apex of a cone, and were all together mortised into a large piece of beechwood, and secured, for the present, with ropes, in a temporary manner. During the short period of one tide all that could further be done for their security was to put a single screw-bolt through the great kneed bats or stanchions on each side of the beams, and screw the nut home.

In this manner these four principal beams were erected, and left in a pretty secure state. The men had commenced while there was about two or three feet of water upon the side of the beacon, and as the sea was smooth they continued the work equally long during flood-tide. Two of the boats being left at the rock to take off the joiners, who were busily employed on the upper parts till two o'clock p.m., this tide's work may be said to have continued for about seven hours, which was the longest that had hitherto been got upon the rock by at least three hours.

When the first boats left the rock with the artificers employed on the lower part of the work during the flood-tide, the beacon had quite a novel appearance. The beams erected formed a common base of about thirty-three feet, meeting at the top, which was about forty-five feet above the rock, and here half a dozen of the artificers were still at work. After clearing the rock the boats made a stop, when three hearty cheers were given, which were returned with equal good-will by those upon the beacon, from the personal interest which every one felt in the prosperity

1807 of this work, so intimately connected with his safety.

All hands having returned to their respective ships, they got a shift of dry clothes and some refreshment. Being Sunday, they were afterwards convened by signal on board of the Lighthouse yacht, when prayers were read; for every heart upon this occasion felt gladness, and every mind was disposed to be thankful for the happy and successful termination of the operations of this day.

Monday, 21st Sept.

The remaining two principal beams were erected in the course of this tide, which, with the assistance of those set up yesterday, was found to be a very simple operation.

The six principal beams of the beacon were thus secured, at least in a temporary manner, in the course of two tides, or in the short space of about eleven hours and a half. Such is the progress that may be made when active hands and willing minds set properly to work in operations of this kind. Having now got the weighty part of this work over, and being thereby relieved of the difficulty both of landing and victualling such a number of men, the *Smeaton* could now be spared, and she was accordingly despatched to Arbroath for a supply of water and provisions, and carried with her six of the artificers who could best be spared.

Tuesday, 22nd Sept.

Wednesday, 23rd Sept.

In going out of the eastern harbour, the boat which the writer steered shipped a sea, that filled her about one-third with water. She had also been hid for a short time, by the waves breaking upon the rock, from the sight of the crew of the preceding boat, who were much alarmed for our safety, imagining for a time that she had gone down.

The *Smeaton* returned from Arbroath this afternoon,

1807

but there was so much sea that she could not be made fast to her moorings, and the vessel was obliged to return to Arbroath without being able either to deliver the provisions or take the artificers on board. The Lighthouse yacht was also soon obliged to follow her example, as the sea was breaking heavily over her bows. After getting two reefs in the mainsail, and the third or storm-jib set, the wind being S.W., she bent to windward, though blowing a hard gale, and got into St. Andrews Bay, where we passed the night under the lee of Fifeness.

Thursday, 24th Sept.

At two o'clock this morning we were in St. Andrews Bay, standing off and on shore, with strong gales of wind at S.W.; at seven we were off the entrance of the Tay; at eight stood towards the rock, and at ten passed to leeward of it, but could not attempt a landing. The beacon, however, appeared to remain in good order, and by six p.m. the vessel had again beaten up to St. Andrews Bay, and got into somewhat smoother water for the night.

Friday, 25th Sept.

At seven o'clock bore away for the Bell Rock, but finding a heavy sea running on it were unable to land. The writer, however, had the satisfaction to observe, with his telescope, that everything about the beacon appeared entire: and although the sea had a most frightful appearance, yet it was the opinion of every one that, since the erection of the beacon, the Bell Rock was divested of many of its terrors, and had it been possible to have got the boats hoisted out and manned, it might have even been found practicable to land. At six it blew so hard that it was found necessary to strike the topmast and take in a third reef of the mainsail, and under this low canvas we soon reached St. Andrews Bay, and got again under the lee of the land for the night. The artificers,

1807 being sea-hardy, were quite reconciled to their quarters on board of the Lighthouse yacht; but it is believed that hardly any consideration would have induced them again to take up their abode in the floating light.

Saturday, 26th Sept.

At daylight the yacht steered towards the Bell Rock, and at eight a.m. made fast to her moorings; at ten, all hands, to the amount of thirty, landed, when the writer had the happiness to find that the beacon had withstood the violence of the gale and the heavy breach of sea, everything being found in the same state in which it had been left on the 21st. The artificers were now enabled to work upon the rock throughout the whole day, both at low and high water, but it required the strictest attention to the state of the weather, in case of their being overtaken with a gale, which might prevent the possibility of getting them off the rock.

Two somewhat memorable circumstances in the annals of the Bell Rock attended the operations of this day: one was the removal of Mr. James Dove, the foreman smith, with his apparatus, from the rock to the upper part of the beacon, where the forge was now erected on a temporary platform, laid on the cross beams or upper framing. The other was the artificers having dined for the first time upon the rock, their dinner being cooked on board of the yacht, and sent to them by one of the boats. But what afforded the greatest happiness and relief was the removal of the large bellows, which had all along been a source of much trouble and perplexity, by their hampering and incommoding the boat which carried the smiths and their apparatus.

Saturday, 3rd Oct.

The wind being west to-day, the weather was very favourable for operations at the rock, and during the

1807

morning and evening tides, with the aid of torchlight, the masons had seven hours' work upon the site of the building. The smiths and joiners, who landed at half-past six a.m., did not leave the rock till a quarter-past eleven p.m., having been at work, with little intermission, for sixteen hours and three-quarters. When the water left the rock, they were employed at the lower parts of the beacon, and as the tide rose or fell, they shifted the place of their operations. From these exertions, the fixing and securing of the beacon made rapid advancement, as the men were now landed in the morning and remained throughout the day. But, as a sudden change of weather might have prevented their being taken off at the proper time of tide, a quantity of bread and water was always kept on the beacon.

During this period of working at the beacon all the day, and often a great part of the night, the writer was much on board of the tender; but, while the masons could work on the rock, and frequently also while it was covered by the tide, he remained on the beacon; especially during the night, as he made a point of being on the rock to the latest hour, and was generally the last person who stepped into the boat. He had laid this down as part of his plan of procedure; and in this way had acquired, in the course of the first season, a pretty complete knowledge and experience of what could actually be done at the Bell Rock, under all circumstances of the weather. By this means also his assistants, and the artificers and mariners, got into a systematic habit of proceeding at the commencement of the work, which, it is believed, continued throughout the whole of the operations.

Sunday 4th Oct.

The external part of the beacon was now finished, with its supports and bracing-chains, and whatever else

1807 was considered necessary for its stability, in so far as the season would permit; and although much was still wanting to complete this fabric, yet it was in such a state that it could be left without much fear of the consequences of a storm. The painting of the upper part was nearly finished this afternoon; and the *Smeaton* had brought off a quantity of brushwood and other articles, for the purpose of heating or charring the lower part of the principal beams, before being laid over with successive coats of boiling pitch, to the height of from eight to twelve feet, or as high as the rise of spring-tides. A small flagstaff having also been erected to-day, a flag was displayed for the first time from the beacon, by which its perspective effect was greatly improved. On this, as on all like occasions at the Bell Rock, three hearty cheers were given; and the steward served out a dram of rum to all hands, while the Lighthouse yacht, *Smeaton*, and floating light, hoisted their colours in compliment to the erection.

Monday, 5th Oct.

In the afternoon, and just as the tide's work was over, Mr. John Rennie, engineer, accompanied by his son Mr. George, on their way to the harbour works of Fraserburgh, in Aberdeenshire, paid a visit to the Bell Rock, in a boat from Arbroath. It being then too late in the tide for landing, they remained on board of the Lighthouse yacht all night, when the writer, who had now been secluded from society for several weeks, enjoyed much of Mr. Rennie's interesting conversation, both on general topics, and professionally upon the progress of the Bell Rock works, on which he was consulted as chief engineer.

Tuesday, 6th Oct.

The artificers landed this morning at nine, after which one of the boats returned to the ship for the writer and Messrs. Rennie, who, upon landing, were saluted with a display of the colours from the beacon and by three cheers from the workmen. Everything was now in a prepared state for leaving the rock, and giving up the works afloat for this season, excepting some small articles, which would still occupy the smiths and joiners for a few days longer. They accordingly shifted on board of the *Smeaton*, while the yacht left the rock for Arbroath, with Messrs. Rennie, the writer, and the remainder of the artificers. But, before taking leave, the steward served out a farewell glass, when three hearty cheers were given, and an earnest wish expressed that everything, in the spring of 1808, might be found in the same state of good order as it was now about to be left.

1807

II

OPERATIONS OF 1808

The writer sailed from Arbroath at one a.m. in the Lighthouse yacht. At seven the floating light was hailed, and all on board found to be well. The crew were observed to have a very healthy-like appearance, and looked better than at the close of the works upon the rock. They seemed only to regret one thing, which was the secession of their cook, Thomas Elliot—not on account of his professional skill, but for his facetious and curious manner. Elliot had something peculiar in his history, and was reported by his comrades to have seen better days. He was, however, happy with his situation on board of the floating light, and, having a taste for music, dancing, and acting plays, he contributed much to the amusement of the ship's company in their dreary abode during the winter months. He had also recommended himself to

1808
Monday, 29th Feb.

1808 their notice as a good shipkeeper, for as it did not answer Elliot to go often ashore, he had always given up his turn of leave to his neighbours. At his own desire he was at length paid off, when he had a considerable balance of wages to receive, which he said would be sufficient to carry him to the West Indies, and he accordingly took leave of the Lighthouse service.

Tuesday, 1st March. At daybreak the Lighthouse yacht, attended by a boat from the floating light, again stood towards the Bell Rock. The weather felt extremely cold this morning, the thermometer being at 34 degrees, with the wind at east, accompanied by occasional showers of snow, and the marine barometer indicated 29·80. At half-past seven the sea ran with such force upon the rock that it seemed doubtful if a landing could be effected. At half-past eight, when it was fairly above water, the writer took his place in the floating light's boat with the artificers, while the yacht's boat followed, according to the general rule of having two boats afloat in landing expeditions of this kind, that, in case of accident to one boat, the other might assist. In several unsuccessful attempts the boats were beat back by the breach of the sea upon the rock. On the eastern side it separated into two distinct waves, which came with a sweep round to the western side, where they met; and at the instance of their confluence the water rose in spray to a considerable height. Watching what the sailors term a *smooth,* we caught a favourable opportunity, and in a very dexterous manner the boats were rowed between the two seas, and made a favourable landing at the western creek.

At the latter end of last season, as was formerly noticed, the beacon was painted white, and from the bleaching of the weather and the sprays of the sea the upper parts were kept clean; but within the range of the tide the principal beams were observed to be thickly coated with a green stuff, the *conferva* of botanists. Notwithstanding the intrusion of these works, which had formerly banished the numerous seals that played about the rock, they were now seen in great numbers, having been in an almost undisturbed state for six months. It had now also, for the first time, got some inhabitants of the feathered tribe: in particular the scarth or cormorant, and the large herring-gull, had made the beacon a resting-place, from its vicinity to their fishing-grounds. About a dozen of these birds had rested upon the cross-beams, which, in some places, were coated with their dung; and their flight, as the boats approached, was a very unlooked-for indication of life and habitation on the Bell Rock, conveying the momentary idea of the conversion of this fatal rock, from being a terror to the mariner, into a residence of man and a safeguard to shipping.

Upon narrowly examining the great iron stanchions with which the beams were fixed to the rock, the writer had the satisfaction of finding that there was not the least appearance of working or shifting at any of the joints or places of connection; and, excepting the loosening of the bracing-chains, everything was found in the same entire state in which it had been left in the month of October. This, in the estimation of the writer, was a matter of no small importance to the future success of the work. He from that moment saw the practicability and propriety of fitting up the beacon, not only as a place of refuge in case of accident to the boats in landing, but as a residence for the artificers during the working months.

While upon the top of the beacon the writer was

1808 reminded by the landing-master that the sea was running high, and that it would be necessary to set off while the rock afforded anything like shelter to the boats, which by this time had been made fast by a long line to the beacon, and rode with much agitation, each requiring two men with boat-hooks to keep them from striking each other, or from ranging up against the beacon. But even under these circumstances the greatest confidence was felt by every one, from the security afforded by this temporary erection. For, supposing the wind had suddenly increased to a gale, and that it had been found unadvisable to go into the boats; or, supposing they had drifted or sprung a leak from striking upon the rocks; in any of these possible and not at all improbable cases, those who might thus have been left upon the rock had now something to lay hold of, and, though occupying this dreary habitation of the sea-gull and the cormorant, affording only bread and water, yet *life* would be preserved, and the mind would still be supported by the hope of being ultimately relieved.

Wednesday, 25th May.

On the 25th of May the writer embarked at Arbroath, on board of the *Sir Joseph Banks,* for the Bell Rock, accompanied by Mr. Logan senior, foreman builder, with twelve masons and two smiths, together with thirteen seamen, including the master, mate, and steward.

Thursday, 26th May.

Mr. James Wilson, now commander of the *Pharos,* floating light, and landing-master, in the room of Mr. Sinclair, who had left the service, came into the writer's cabin this morning at six o'clock, and intimated that there was a good appearance of landing on the rock. Everything being arranged, both boats proceeded in company, and at eight a.m. they reached the rock. The lighthouse colours were immediately

1808

hoisted upon the flagstaff of the beacon, a compliment which was duly returned by the tender and floating light, when three hearty cheers were given, and a glass of rum was served out to all hands to drink success to the operations of 1808.

Friday, 27th May.

This morning the wind was at east, blowing a fresh gale, the weather being hazy, with a considerable breach of sea setting in upon the rock. The morning bell was therefore rung, in some doubt as to the practicability of making a landing. After allowing the rock to get fully up, or to be sufficiently left by the tide, that the boats might have some shelter from the range of the sea, they proceeded at 8 a.m., and upon the whole made a pretty good landing; and after two hours and three-quarters' work returned to the ship in safety.

In the afternoon the wind considerably increased, and, as a pretty heavy sea was still running, the tender rode very hard, when Mr. Taylor, the commander, found it necessary to take in the bowsprit, and strike the fore and main topmasts, that she might ride more easily. After consulting about the state of the weather, it was resolved to leave the artificers on board this evening, and carry only the smiths to the rock, as the sharpening of the irons was rather behind, from their being so much broken and blunted by the hard and tough nature of the rock, which became much more compact and hard as the depth of excavation was increased. Besides avoiding the risk of encumbering the boats with a number of men who had not yet got the full command of the oar in a breach of sea, the writer had another motive for leaving them behind. He wanted to examine the site of the building without interruption, and to take the comparative levels of the different inequalities of

1808 its area; and as it would have been painful to have seen men standing idle upon the Bell Rock, where all moved with activity, it was judged better to leave them on board. The boats landed at half-past seven p.m., and the landing-master, with the seamen, was employed during this tide in cutting the seaweeds from the several paths leading to the landing-places, to render walking more safe, for, from the slippery state of the surface of the rock, many severe tumbles had taken place. In the meantime the writer took the necessary levels, and having carefully examined the site of the building and considered all its parts, it still appeared to be necessary to excavate to the average depth of fourteen inches over the whole area of the foundation.

Saturday, 28th May.

The wind still continued from the eastward with a heavy swell; and to-day it was accompanied with foggy weather and occasional showers of rain. Notwithstanding this, such was the confidence which the erection of the beacon had inspired that the boats landed the artificers on the rock under very unpromising circumstances, at half-past eight, and they continued at work till half-past eleven, being a period of three hours, which was considered a great tide's work in the present low state of the foundation. Three of the masons on board were so afflicted with sea-sickness that they had not been able to take any food for almost three days, and they were literally assisted into the boats this morning by their companions. It was, however, not a little surprising to see how speedily these men revived upon landing on the rock and eating a little dulse. Two of them afterwards assisted the sailors in collecting the chips of stone and carrying them out of the way of the pickmen; but the third complained of a pain in his head, and was still unable

1808

to do anything. Instead of returning to the tender with the boats, these three men remained on the beacon all day, and had their victuals sent to them along with the smiths'. From Mr. Dove, the foreman smith, they had much sympathy, for he preferred remaining on the beacon at all hazards, to be himself relieved from the malady of sea-sickness. The wind continuing high, with a heavy sea, and the tide falling late, it was not judged proper to land the artificers this evening, but in the twilight the boats were sent to fetch the people on board who had been left on the rock.

Sunday, 29th May.

The wind was from the S.W. to-day, and the signal-bell rung, as usual, about an hour before the period for landing on the rock. The writer was rather surprised, however, to hear the landing-master repeatedly call, 'All hands for the rock!' and, coming on deck, he was disappointed to find the seamen only in the boats. Upon inquiry, it appeared that some misunderstanding had taken place about the wages of the artificers for Sundays. They had preferred wages for seven days statedly to the former mode of allowing a day for each tide's work on Sunday, as they did not like the appearance of working for double or even treble wages on Sunday, and would rather have it understood that their work on that day arose more from the urgency of the case than with a view to emolument. This having been judged creditable to their religious feelings, and readily adjusted to their wish, the boats proceeded to the rock, and the work commenced at nine a.m.

Monday, 30th May.

Mr. Francis Watt commenced, with five joiners, to fit up a temporary platform upon the beacon, about twenty-five feet above the highest part of the rock. This platform was to be used as the site of the smith's

1808 forge, after the beacon should be fitted up as a barrack; and here also the mortar was to be mixed and prepared for the building, and it was accordingly termed the Mortar Gallery.

The landing-master's crew completed the discharging from the *Smeaton* of her cargo of the cast-iron rails and timber. It must not here be omitted to notice that the *Smeaton* took in ballast from the Bell Rock, consisting of the shivers or chips of stone produced by the workmen in preparing the site of the building, which were now accumulating in great quantities on the rock. These the boats loaded, after discharging the iron. The object in carrying off these chips, besides ballasting the vessel, was to get them permanently out of the way, as they were apt to shift about from place to place with every gale of wind; and it often required a considerable time to clear the foundation a second time of this rubbish. The circumstance of ballasting a ship at the Bell Rock afforded great entertainment, especially to the sailors; and it was perhaps with truth remarked that the *Smeaton* was the first vessel that had ever taken on board ballast at the Bell Rock. Mr. Pool, the commander of this vessel, afterwards acquainted the writer that, when the ballast was landed upon the quay at Leith, many persons carried away specimens of it, as part of a cargo from the Bell Rock; when he added, that such was the interest excited, from the number of specimens carried away, that some of his friends suggested that he should have sent the whole to the Cross of Edinburgh, where each piece might have sold for a penny.

Tuesday, 31st May. In the evening the boats went to the rock, and brought the joiners and smiths, and their sickly companions, on board of the tender. These also brought

1808

with them two baskets full of fish, which they had caught at high-water from the beacon, reporting, at the same time, to their comrades, that the fish were swimming in such numbers over the rock at high-water that it was completely hid from their sight, and nothing seen but the movement of thousands of fish. They were almost exclusively of the species called the podlie, or young coal-fish. This discovery, made for the first time to-day by the workmen, was considered fortunate, as an additional circumstance likely to produce an inclination among the artificers to take up their residence in the beacon, when it came to be fitted up as a barrack.

Tuesday, 7th June.

At three o'clock in the morning the ship's bell was rung as the signal for landing at the rock. When the landing was to be made before breakfast, it was customary to give each of the artificers and seamen a dram and a biscuit, and coffee was prepared by the steward for the cabins. Exactly at four o'clock the whole party landed from three boats, including one of those belonging to the floating light, with a part of that ship's crew, which always attended the works in moderate weather. The landing-master's boat, called the *Seaman*, but more commonly called the *Lifeboat*, took the lead. The next boat, called the *Mason*, was generally steered by the writer; while the floating light's boat, *Pharos*, was under the management of the boatswain of that ship.

Having now so considerable a party of workmen and sailors on the rock, it may be proper here to notice how their labours were directed. Preparations having been made last month for the erection of a second forge upon the beacon, the smiths commenced their operations both upon the lower and higher platforms. They were employed in sharpening the

1808 picks and irons for the masons, and in making bats and other apparatus of various descriptions connected with the fitting of the railways. The landing-master's crew were occupied in assisting the millwrights in laying the railways to hand. Sailors, of all other descriptions of men, are the most accommodating in the use of their hands. They worked freely with the boring-irons, and assisted in all the operations of the railways, acting by turns as boatmen, seamen, and artificers. We had no such character on the Bell Rock as the common labourer. All the operations of this department were cheerfully undertaken by the seamen, who, both on the rock and on shipboard, were the inseparable companions of every work connected with the erection of the Bell Rock Lighthouse. It will naturally be supposed that about twenty-five masons, occupied with their picks in executing and preparing the foundation of the lighthouse, in the course of a tide of about three hours, would make a considerable impression upon an area even of forty-two feet in diameter. But in proportion as the foundation was deepened, the rock was found to be much more hard and difficult to work, while the baling and pumping of water became much more troublesome. A joiner was kept almost constantly employed in fitting the picks to their handles, which, as well as the points to the irons, were very frequently broken.

The Bell Rock this morning presented by far the most busy and active appearance it had exhibited since the erection of the principal beams of the beacon. The surface of the rock was crowded with men, the two forges flaming, the one above the other, upon the beacon, while the anvils thundered with the rebounding noise of their wooden supports, and formed a curious contrast with the occasional clamour of the surges. The wind was westerly, and the weather being extremely agreeable, as soon after breakfast as the tide had sufficiently overflowed the rock to float the boats over it, the smiths, with a number of the artificers, returned to the beacon, carrying their fishing-tackle along with them. In the course of the forenoon, the beacon exhibited a still more extraordinary appearance than the rock had done in the morning. The sea being smooth, it seemed to be afloat upon the water, with a number of men supporting themselves in all the variety of attitude and position: while, from the upper part of this wooden house, the volumes of smoke which ascended from the forges gave the whole a very curious and fanciful appearance.

In the course of this tide it was observed that a heavy swell was setting in from the eastward, and the appearance of the sky indicated a change of weather, while the wind was shifting about. The barometer also had fallen from 30 in. to 29·6. It was, therefore, judged prudent to shift the vessel to the S.W. or more distant buoy. Her bowsprit was also soon afterwards taken in, the topmasts struck, and everything made *snug*, as seamen term it, for a gale. During the course of the night the wind increased and shifted to the eastward, when the vessel rolled very hard, and the sea often broke over her bows with great force.

Wednesday, 8th June.

Although the motion of the tender was much less than that of the floating light—at least, in regard to the rolling motion—yet she *sended*, or pitched, much. Being also of a very handsome build, and what seamen term very *clean aft*, the sea often struck the counter with such force that the writer, who possessed the aftermost cabin, being unaccustomed to this new vessel, could not divest himself of uneasiness; for

1808 when her stern fell into the sea, it struck with so much violence as to be more like the resistance of a rock than the sea. The water, at the same time, often rushed with great force up the rudder-case, and, forcing up the valve of the water-closet, the floor of his cabin was at times laid under water. The gale continued to increase, and the vessel rolled and pitched in such a manner that the hawser by which the tender was made fast to the buoy snapped, and she went adrift. In the act of swinging round to the wind she shipped a very heavy sea, which greatly alarmed the artificers, who imagined that we had got upon the rock; but this, from the direction of the wind, was impossible. The writer, however, sprung upon deck, where he found the sailors busily employed in rigging out the bowsprit and in setting sail. From the easterly direction of the wind, it was considered most advisable to steer for the Firth of Forth, and there wait a change of weather. At two p.m. we accordingly passed the Isle of May, at six anchored in Leith Roads, and at eight the writer landed, when he came in upon his friends, who were not a little surprised at his unexpected appearance, which gave an instantaneous alarm for the safety of things at the Bell Rock.

Thursday, 9th June.

The wind still continued to blow very hard at E. by N., and the *Sir Joseph Banks* rode heavily, and even drifted with both anchors ahead, in Leith Roads. The artificers did not attempt to leave the ship last night; but there being upwards of fifty people on board, and the decks greatly lumbered with the two large boats, they were in a very crowded and impatient state on board. But to-day they got ashore, and amused themselves by walking about the streets of Edinburgh, some in very humble apparel, from having only the worst of their jackets with them, which, though quite suitable for their work, were hardly fit for public inspection, being not only tattered, but greatly stained with the red colour of the rock.

1808

Friday, 10th June.

To-day the wind was at S.E., with light breezes and foggy weather. At six a.m. the writer again embarked for the Bell Rock, when the vessel immediately sailed. At eleven p.m., there being no wind, the kedge-anchor was *let go* off Anstruther, one of the numerous towns on the coast of Fife, where we waited the return of the tide.

Saturday, 11th June.

At six a.m. the *Sir Joseph* got under weigh, and at eleven was again made fast to the southern buoy at the Bell Rock. Though it was now late in the tide, the writer, being anxious to ascertain the state of things after the gale, landed with the artificers to the number of forty-four. Everything was found in an entire state; but, as the tide was nearly gone, only half an hour's work had been got when the site of the building was overflowed. In the evening the boats again landed at nine, and after a good tide's work of three hours with torchlight, the work was left off at midnight. To the distant shipping the appearance of things under night on the Bell Rock, when the work was going forward, must have been very remarkable, especially to those who were strangers to the operations. Mr. John Reid, principal lightkeeper, who also acted as master of the floating light during the working months at the rock, described the appearance of numerous lights situated so low in the water, when seen at the distance of two or three miles, as putting him in mind of Milton's description of the fiends in the lower regions, adding, 'for it seems greatly to surpass Will-o'-the-Wisp, or any of those earthly spectres of which we have so often heard.'

1808
Monday, 13th June.

From the difficulties attending the landing on the rock, owing to the breach of sea which had for days past been around it, the artificers showed some backwardness at getting into the boats this morning; but after a little explanation this was got over. It was always observable that for some time after anything like danger had occurred at the rock, the workmen became much more cautious, and on some occasions their timidity was rather troublesome. It fortunately happened, however, that along with the writer's assistants and the sailors there were also some of the artificers themselves who felt no such scruples, and in this way these difficulties were the more easily surmounted. In matters where life is in danger it becomes necessary to treat even unfounded prejudices with tenderness, as an accident, under certain circumstances, would not only have been particularly painful to those giving directions, but have proved highly detrimental to the work, especially in the early stages of its advancement.

At four o'clock fifty-eight persons landed; but the tides being extremely languid, the water only left the higher parts of the rock, and no work could be done at the site of the building. A third forge was, however, put in operation during a short time, for the greater conveniency of sharpening the picks and irons, and for purposes connected with the preparations for fixing the railways on the rock. The weather towards the evening became thick and foggy, and there was hardly a breath of wind to ruffle the surface of the water. Had it not, therefore, been for the noise from the anvils of the smiths who had been left on the beacon throughout the day, which afforded a guide for the boats, a landing could not have been attempted this evening, especially with such a com-

1808

pany of artificers. This circumstance confirmed the writer's opinion with regard to the propriety of connecting large bells to be rung with machinery in the lighthouse, to be tolled day and night during the continuance of foggy weather.

Thursday, 23rd June.

The boats landed this evening, when the artificers had again two hours' work. The weather still continuing very thick and foggy, more difficulty was experienced in getting on board of the vessels to-night than had occurred on any previous occasion, owing to a light breeze of wind which carried the sound of the bell, and the other signals made on board of the vessels, away from the rock. Having fortunately made out the position of the sloop *Smeaton* at the N.E. buoy—to which we were much assisted by the barking of the ship's dog,—we parted with the *Smeaton's* boat, when the boats of the tender took a fresh departure for that vessel, which lay about half a mile to the south-westward. Yet such is the very deceiving state of the tides, that, although there was a small binnacle and compass in the landing-master's boat, we had, nevertheless, passed the *Sir Joseph* a good way, when, fortunately, one of the sailors catched the sound of a blowing-horn. The only fire-arms on board were a pair of swivels of one-inch calibre; but it is quite surprising how much the sound is lost in foggy weather, as the report was heard but at a very short distance. The sound from the explosion of gunpowder is so instantaneous that the effect of the small guns was not so good as either the blowing of a horn or the tolling of a bell, which afforded a more constant and steady direction for the pilot.

Wednesday, 6th July.

Landed on the rock with the three boats belonging to the tender at five p.m., and began immediately to

1808 bale the water out of the foundation-pit with a number of buckets, while the pumps were also kept in action with relays of artificers and seamen. The work commenced upon the higher parts of the foundation as the water left them, but it was now pretty generally reduced to a level. About twenty men could be conveniently employed at each pump, and it is quite astonishing in how short a time so great a body of water could be drawn off. The water in the foundation-pit at this time measured about two feet in depth, on an area of forty-two feet in diameter, and yet it was drawn off in the course of about half an hour. After this the artificers commenced with their picks and continued at work for two hours and a half, some of the sailors being at the same time busily employed in clearing the foundation of chips and in conveying the irons to and from the smiths on the beacon, where they were sharped. At eight o'clock the sea broke in upon us and overflowed the foundation-pit, when the boats returned to the tender.

Thursday, 7th July.

The landing-master's bell rung this morning about four o'clock, and at half-past five, the foundation being cleared, the work commenced on the site of the building. But from the moment of landing, the squad of joiners and millwrights was at work upon the higher parts of the rock in laying the railways, while the anvils of the smith resounded on the beacon, and such columns of smoke ascended from the forges that they were often mistaken by strangers at a distance for a ship on fire. After continuing three hours at work the foundation of the building was again overflowed, and the boats returned to the ship at half-past eight o'clock. The masons and pickmen had, at this period, a pretty long day on board of the tender, but the smiths and joiners were kept constantly at work

1808

upon the beacon, the stability and great conveniency of which had now been so fully shown that no doubt remained as to the propriety of fitting it up as a barrack. The workmen were accordingly employed, during the period of high-water, in making preparations for this purpose.

The foundation-pit now assumed the appearance of a great platform, and the late tides had been so favourable that it became apparent that the first course, consisting of a few irregular and detached stones for making up certain inequalities in the interior parts of the site of the building, might be laid in the course of the present spring-tides. Having been enabled to-day to get the dimensions of the foundation, or first stone, accurately taken, a mould was made of its figure, when the writer left the rock, after the tide's work of this morning, in a fast rowing-boat for Arbroath; and, upon landing, two men were immediately set to work upon one of the blocks from Mylnefield quarry, which was prepared in the course of the following day, as the stone-cutters relieved each other, and worked both night and day, so that it was sent off in one of the stone-lighters without delay.

Saturday, 9th July.

The site of the foundation-stone was very difficult to work, from its depth in the rock; but being now nearly prepared, it formed a very agreeable kind of pastime at high-water for all hands to land the stone itself upon the rock. The landing-master's crew and artificers accordingly entered with great spirit into this operation. The stone was placed upon the deck of the *Hedderwick* praam-boat, which had just been brought from Leith, and was decorated with colours for the occasion. Flags were also displayed from the shipping in the offing, and upon the beacon. Here the writer took his station with the greater part of the artificers, who supported

1808 themselves in every possible position while the boats towed the praam from her moorings and brought her immediately over the site of the building, where her grappling anchors were let go. The stone was then lifted off the deck by a tackle hooked into a Lewis bat inserted into it, when it was gently lowered into the water and grounded on the site of the building, amidst the cheering acclamations of about sixty persons.

Sunday, 10th July. At eleven o'clock the foundation-stone was laid to hand. It was of a square form, containing about twenty cubic feet, and had the figures, or date, of 1808 simply cut upon it with a chisel. A derrick, or spar of timber, having been erected at the edge of the hole and guyed with ropes, the stone was then hooked to the tackle and lowered into its place, when the writer, attended by his assistants—Mr. Peter Logan, Mr. Francis Watt, and Mr. James Wilson,—applied the square, the level, and the mallet, and pronounced the following benediction: 'May the great Architect of the Universe complete and bless this building,' on which three hearty cheers were given, and success to the future operations was drunk with the greatest enthusiasm.

Tuesday, 26th July. The wind being at S.E. this evening, we had a pretty heavy swell of sea upon the rock, and some difficulty attended our getting off in safety, as the boats got aground in the creek and were in danger of being upset. Upon extinguishing the torchlights, about twelve in number, the darkness of the night seemed quite horrible; the water being also much charged with the phosphorescent appearance which is familiar to every one on shipboard, the waves, as they dashed upon the rock, were in some degree like so much liquid flame. The scene, upon the whole, was truly awful!

Wednesday 27th July. In leaving the rock this evening everything, after the torches were extinguished, had the same dismal

1808

appearance as last night, but so perfectly acquainted were the landing-master and his crew with the position of things at the rock, that comparatively little inconveniency was experienced on these occasions when the weather was moderate; such is the effect of habit, even in the most unpleasant situations. If, for example, it had been proposed to a person accustomed to a city life, at once to take up his quarters off a sunken reef and land upon it in boats at all hours of the night, the proposition must have appeared quite impracticable and extravagant; but this practice coming progressively upon the artificers, it was ultimately undertaken with the greatest alacrity. Notwithstanding this, however, it must be acknowledged that it was not till after much labour and peril, and many an anxious hour, that the writer is enabled to state that the site of the Bell Rock Lighthouse is fully prepared for the first entire course of the building.

Friday, 12th Aug.

The artificers landed this morning at half-past ten, and after an hour and a half's work eight stones were laid, which completed the first entire course of the building, consisting of 123 blocks, the last of which was laid with three hearty cheers.

Saturday, 10th Sept.

Landed at nine a.m., and by a quarter-past twelve noon twenty-three stones had been laid. The works being now somewhat elevated by the lower courses, we got quit of the very serious inconvenience of pumping water to clear the foundation-pit. This gave much facility to the operations, and was noticed with expressions of as much happiness by the artificers as the seamen had shown when relieved of the continual trouble of carrying the smith's bellows off the rock prior to the erection of the beacon.

Wednesday, 21st Sept.

Mr. Thomas Macurich, mate of the *Smeaton*, and James Scott, one of the crew, a young man about

1808 eighteen years of age, immediately went into their boat to make fast a hawser to the ring in the top of the floating buoy of the moorings, and were forthwith to proceed to land their cargo, so much wanted, at the rock. The tides at this period were very strong, and the mooring-chain, when sweeping the ground, had caught hold of a rock or piece of wreck by which the chain was so shortened that when the tide flowed the buoy got almost under water, and little more than the ring appeared at the surface. When Macurich and Scott were in the act of making the hawser fast to the ring, the chain got suddenly disentangled at the bottom, and this large buoy, measuring about seven feet in height and three feet in diameter at the middle, tapering to both ends, being what seamen term a *Nun-buoy,* vaulted or sprung up with such force that it upset the boat, which instantly filled with water. Mr. Macurich, with much exertion, succeeded in getting hold of the boat's gunwale, still above the surface of the water, and by this means was saved; but the young man Scott was unfortunately drowned. He had in all probability been struck about the head by the ring of the buoy, for although surrounded with the oars and the thwarts of the boat which floated near him, yet he seemed entirely to want the power of availing himself of such assistance, and appeared to be quite insensible, while Pool, the master of the *Smeaton,* called loudly to him; and before assistance could be got from the tender, he was carried away by the strength of the current and disappeared.

The young man Scott was a great favourite in the service, having had something uncommonly mild and complaisant in his manner; and his loss was therefore universally regretted. The circumstances of his case were also peculiarly distressing to his mother, as her husband, who was a seaman, had for three years past been confined to a French prison, and the deceased was the chief support of the family. In order in some measure to make up the loss to the poor woman for the monthly aliment regularly allowed her by her late son, it was suggested that a younger boy, a brother of the deceased, might be taken into the service. This appeared to be rather a delicate proposition, but it was left to the landing-master to arrange according to circumstances; such was the resignation, and at the same time the spirit, of the poor woman, that she readily accepted the proposal, and in a few days the younger Scott was actually afloat in the place of his brother. On representing this distressing case to the Board, the Commissioners were pleased to grant an annuity of £5 to Scott's mother.

1808

The *Smeaton*, not having been made fast to the buoy, had, with the ebb-tide, drifted to leeward a considerable way eastward of the rock, and could not, till the return of the flood-tide, be worked up to her moorings, so that the present tide was lost, notwithstanding all exertions which had been made both ashore and afloat with this cargo. The artificers landed at six a.m.; but, as no materials could be got upon the rock this morning, they were employed in boring trenail holes and in various other operations, and after four hours' work they returned on board the tender. When the *Smeaton* got up to her moorings, the landing-master's crew immediately began to unload her. There being too much wind for towing the praams in the usual way, they were warped to the rock in the most laborious manner by their windlasses, with successive grapplings and hawsers laid out for this purpose. At six p.m. the artificers landed, and continued at work till half-past ten, when the remaining seventeen stones were

1808 laid which completed the third entire course, or fourth of the lighthouse, with which the building operations were closed for the season.

III

OPERATIONS OF 1809

1809 Wednesday, 24th May. The last night was the first that the writer had passed in his old quarters on board of the floating light for about twelve months, when the weather was so fine and the sea so smooth that even here he felt but little or no motion, excepting at the turn of the tide, when the vessel gets into what the seamen term the *trough of the sea.* At six a.m. Mr. Watt, who conducted the operations of the railways and beacon-house, had landed with nine artificers. At half-past one p.m. Mr. Peter Logan had also landed with fifteen masons, and immediately proceeded to set up the crane. The sheer-crane or apparatus for lifting the stones out of the praam-boats at the eastern creek had been already erected, and the railways now formed about two-thirds of an entire circle round the building: some progress had likewise been made with the reach towards the western landing-place. The floors being laid, the beacon now assumed the appearance of a habitation. The *Smeaton* was at her moorings, with the *Fernie* praam-boat astern, for which she was laying down moorings, and the tender being also at her station, the Bell Rock had again put on its former busy aspect.

Wednesday, 31st May. The landing-master's bell, often no very favourite sound, rung at six this morning; but on this occasion, it is believed, it was gladly received by all on board, as the welcome signal of the return of better weather.

1809

The masons laid thirteen stones to-day, which the seamen had landed, together with other building materials. During these twenty-four hours the wind was from the south, blowing fresh breezes, accompanied with showers of snow. In the morning the snow showers were so thick that it was with difficulty the landing-master, who always steered the leading boat, could make his way to the rock through the drift. But at the Bell Rock neither snow nor rain, nor fog nor wind, retarded the progress of the work, if unaccompanied by a heavy swell or breach of the sea.

The weather during the months of April and May had been uncommonly boisterous, and so cold that the thermometer seldom exceeded 40°, while the barometer was generally about 29·50. We had not only hail and sleet, but the snow on the last day of May lay on the decks and rigging of the ship to the depth of about three inches; and, although now entering upon the month of June, the length of the day was the chief indication of summer. Yet such is the effect of habit, and such was the expertness of the landing-master's crew, that, even in this description of weather, seldom a tide's work was lost. Such was the ardour and zeal of the heads of the several departments at the rock, including Mr. Peter Logan, foreman builder, Mr. Francis Watt, foreman millwright, and Captain Wilson, landing-master, that it was on no occasion necessary to address them, excepting in the way of precaution or restraint. Under these circumstances, however, the writer not unfrequently felt considerable anxiety, of which this day's experience will afford an example.

Thursday, 1st June.

This morning, at a quarter-past eight, the artificers were landed as usual, and, after three hours and three-

1809 quarters' work, five stones were laid, the greater part of this tide having been taken up in completing the boring and trenailing of the stones formerly laid. At noon the writer, with the seamen and artificers, proceeded to the tender, leaving on the beacon the joiners, and several of those who were troubled with sea-sickness—among whom was Mr. Logan, who remained with Mr. Watt—counting altogether eleven persons. During the first and middle parts of these twenty-four hours the wind was from the east, blowing what the seamen term 'fresh breezes'; but in the afternoon it shifted to E.N.E., accompanied with so heavy a swell of sea that the *Smeaton* and tender struck their topmasts, launched in their bolt-sprits, and 'made all snug' for a gale. At four p.m. the *Smeaton* was obliged to slip her moorings, and passed the tender, drifting before the wind, with only the foresail set. In passing, Mr. Pool hailed that he must run for the Firth of Forth to prevent the vessel from 'riding under.'

On board of the tender the writer's chief concern was about the eleven men left upon the beacon. Directions were accordingly given that everything about the vessel should be put in the best possible state, to present as little resistance to the wind as possible, that she might have the better chance of riding out the gale. Among these preparations the best bower cable was bent, so as to have a second anchor in readiness in case the mooring-hawser should give way, that every means might be used for keeping the vessel within sight of the prisoners on the beacon, and thereby keep them in as good spirits as possible. From the same motive the boats were kept afloat that they might be less in fear of the vessel leaving her station. The landing-master had, however, repeatedly expressed his anxiety for the safety of the boats, and wished much to have them hoisted on board. At seven p.m. one of the boats, as he feared, was unluckily filled with sea from a wave breaking into her, and it was with great difficulty that she could be baled out and got on board, with the loss of her oars, rudder, and loose thwarts. Such was the motion of the ship that in taking this boat on board her gunwale was stove in, and she otherwise received considerable damage. Night approached, but it was still found quite impossible to go near the rock. Consulting, therefore, the safety of the second boat, she also was hoisted on board of the tender.

At this time the cabins of the beacon were only partially covered, and had neither been provided with bedding nor a proper fireplace, while the stock of provisions was but slender. In these uncomfortable circumstances the people on the beacon were left for the night, nor was the situation of those on board of the tender much better. The rolling and pitching motion of the ship was excessive; and, excepting to those who had been accustomed to a residence in the floating light, it seemed quite intolerable. Nothing was heard but the hissing of the winds and the creaking of the bulkheads or partitions of the ship; the night was, therefore, spent in the most unpleasant reflections upon the condition of the people on the beacon, especially in the prospect of the tender being driven from her moorings. But, even in such a case, it afforded some consolation that the stability of the fabric was never doubted, and that the boats of the floating light were at no great distance, and ready to render the people on the rock the earliest assistance which the weather would permit. The writer's cabin

1809 being in the sternmost part of the ship, which had what sailors term a good entry, or was sharp built, the sea, as before noticed, struck her counter with so much violence that the water, with a rushing noise, continually forced its way up the rudder-case, lifted the valve of the water-closet, and overran the cabin floor. In these circumstances daylight was eagerly looked for, and hailed with delight, as well by those afloat as by the artificers upon the rock.

Friday, 2nd June.

In the course of the night the writer held repeated conversations with the officer on watch, who reported that the weather continued much in the same state, and that the barometer still indicated 29·20 inches. At six a.m. the landing-master considered the weather to have somewhat moderated; and, from certain appearances of the sky, he was of opinion that a change for the better would soon take place. He accordingly proposed to attempt a landing at low-water, and either get the people off the rock, or at least ascertain what state they were in. At nine a.m. he left the vessel with a boat well manned, carrying with him a supply of cooked provisions and a tea-kettle full of mulled port wine for the people on the beacon, who had not had any regular diet for about thirty hours, while they were exposed during that period, in a great measure, both to the winds and the sprays of the sea. The boat having succeeded in landing, she returned at eleven a.m. with the artificers, who had got off with considerable difficulty, and who were heartily welcomed by all on board.

Upon inquiry it appeared that three of the stones last laid upon the building had been partially lifted from their beds by the force of the sea, and were now held only by the trenails, and that the cast-iron sheer-crane had again been thrown down and completely broken. With regard to the beacon, the sea at high-water had lifted part of the mortar gallery or lowest floor, and washed away all the lime-casks and other movable articles from it; but the principal parts of this fabric had sustained no damage. On pressing Messrs. Logan and Watt on the situation of things in the course of the night, Mr. Logan emphatically said: 'That the beacon had an *ill-faured*[1] *twist* when the sea broke upon it at high-water, but that they were not very apprehensive of danger.' On inquiring as to how they spent the night, it appeared that they had made shift to keep a small fire burning, and by means of some old sails defended themselves pretty well from the sea sprays.

It was particularly mentioned that by the exertions of James Glen, one of the joiners, a number of articles were saved from being washed off the mortar gallery. Glen was also very useful in keeping up the spirits of the forlorn party. In the early part of life he had undergone many curious adventures at sea, which he now recounted somewhat after the manner of the tales of the *Arabian Nights.* When one observed that the beacon was a most comfortless lodging, Glen would presently introduce some of his exploits and hardships, in comparison with which the state of things at the beacon bore an aspect of comfort and happiness. Looking to their slender stock of provisions, and their perilous and uncertain chance of speedy relief, he would launch out into an account of one of his expeditions in the North Sea, when the vessel, being much disabled in a storm, was driven before the wind with the loss of almost all their provisions; and the ship being much infested with rats, the crew hunted these vermin with great eagerness to help their scanty

[1] Ill-formed—ugly.—[R. L. S.]

1809 allowance. By such means Glen had the address to make his companions, in some measure, satisfied, or at least passive, with regard to their miserable prospects upon this half-tide rock in the middle of the ocean. This incident is noticed, more particularly, to show the effects of such a happy turn of mind, even under the most distressing and ill-fated circumstances.

Saturday, 17th June.

At eight a.m. the artificers and sailors, forty-five in number, landed on the rock, and after four hours' work seven stones were laid. The remainder of this tide, from the threatening appearance of the weather, was occupied in trenailing and making all things as secure as possible. At twelve noon the rock and building were again overflowed, when the masons and seamen went on board of the tender, but Mr. Watt, with his squad of ten men, remained on the beacon throughout the day. As it blew fresh from the N.W. in the evening, it was found impracticable either to land the building artificers or to take the artificers off the beacon, and they were accordingly left there all night, but in circumstances very different from those of the 1st of this month. The house, being now in a more complete state, was provided with bedding, and they spent the night pretty well, though they complained of having been much disturbed at the time of high-water by the shaking and tremulous motion of their house and by the plashing noise of the sea upon the mortar gallery. Here James Glen's versatile powers were again at work in cheering up those who seemed to be alarmed, and in securing everything as far as possible. On this occasion he had only to recall to the recollections of some of them the former night which they had spent on the beacon, the wind and sea being then much higher, and their habitation in a far less comfortable state.

1809

The wind still continuing to blow fresh from the N.W., at five p.m. the writer caused a signal to be made from the tender for the *Smeaton* and *Patriot* to slip their moorings, when they ran for Lunan Bay, an anchorage on the east side of the Redhead. Those on board of the tender spent but a very rough night, and perhaps slept less soundly than their companions on the beacon, especially as the wind was at N.W., which caused the vessel to ride with her stern towards the Bell Rock; so that, in the event of anything giving way, she could hardly have escaped being stranded upon it.

Sunday, 18th June.

The weather having moderated to-day, the wind shifted to the westward. At a quarter-past nine a.m. the artificers landed from the tender and had the pleasure to find their friends who had been left on the rock quite hearty, alleging that the beacon was the preferable quarters of the two.

Saturday, 24th June.

Mr. Peter Logan, the foreman builder, and his squad, twenty-one in number, landed this morning at three o'clock, and continued at work four hours and a quarter, and after laying seventeen stones returned to the tender. At six a.m. Mr. Francis Watt and his squad of twelve men landed, and proceeded with their respective operations at the beacon and railways, and were left on the rock during the whole day without the necessity of having any communication with the tender, the kitchen of the beacon-house being now fitted up. It was to-day, also, that Peter Fortune—a most obliging and well-known character in the Lighthouse service—was removed from the tender to the beacon as cook and steward, with a stock of provisions as ample as his limited store-room would admit.

When as many stones were built as comprised this

1809 day's work, the demand for mortar was proportionally increased, and the task of the mortar-makers on these occasions was both laborious and severe. This operation was chiefly performed by John Watt—a strong, active quarrier by profession,—who was a perfect character in his way, and extremely zealous in his department. While the operations of the mortar-makers continued, the forge upon the gallery was not generally in use; but, as the working hours of the builders extended with the height of the building, the forge could not be so long wanted, and then a sad confusion often ensued upon the circumscribed floor of the mortar gallery, as the operations of Watt and his assistants trenched greatly upon those of the smiths. Under these circumstances the boundary of the smiths was much circumscribed, and they were personally annoyed, especially in blowy weather, with the dust of the lime in its powdered state. The mortar-makers, on the other hand, were often not a little distressed with the heat of the fire and the sparks elicited on the anvil, and not unaptly complained that they were placed between the 'devil and the deep sea.'

Sunday 25th June. The work being now about ten feet in height, admitted of a rope-ladder being distended[1] between the beacon and the building. By this 'Jacob's Ladder,' as the seamen termed it, a communication was kept up with the beacon while the rock was considerably under water. One end of it being furnished with tackle-blocks, was fixed to the beams of the beacon, at the level of the mortar gallery, while the further end was connected with the upper course of the building by means of two Lewis bats which

[1] This is an incurable illusion of my grandfather's; he always writes distended' for 'extended.'—[R. L. S.]

were lifted from course to course as the work advanced. In the same manner a rope furnished with a travelling pulley was distended for the purpose of transporting the mortar-buckets, and other light articles between the beacon and the building, which also proved a great conveniency to the work. At this period the rope-ladder and tackle for the mortar had a descent from the beacon to the building; by and by they were on a level, and towards the end of the season, when the solid part had attained its full height, the ascent was from the mortar gallery to the building.

1809

The artificers landed on the rock this morning at a quarter-past six, and remained at work five hours. The cooking apparatus being now in full operation, all hands had breakfast on the beacon at the usual hour, and remained there throughout the day. The crane upon the building had to be raised to-day from the eighth to the ninth course, an operation which now required all the strength that could be mustered for working the guy-tackles; for as the top of the crane was at this time about thirty-five feet above the rock, it became much more unmanageable. While the beam was in the act of swinging round from one guy to another, a great strain was suddenly brought upon the opposite tackle, with the end of which the artificers had very improperly neglected to take a turn round some stationary object, which would have given them the complete command of the tackle. Owing to this simple omission, the crane got a preponderancy to one side, and fell upon the building with a terrible crash. The surrounding artificers immediately flew in every direction to get out of its way; but Michael Wishart, the principal builder, having unluckily stumbled upon one of the uncut trenails, fell upon his back. His body fortunately got between the

Friday, 30th June.

1809 movable beam and the upright shaft of the crane, and was thus saved; but his feet got entangled with the wheels of the crane and were severely injured. Wishart, being a robust young man, endured his misfortune with wonderful firmness; he was laid upon one of the narrow framed beds of the beacon and despatched in a boat to the tender, where the writer was when this accident happened, not a little alarmed on missing the crane from the top of the building, and at the same time seeing a boat rowing towards the vessel with great speed. When the boat came alongside with poor Wishart, stretched upon a bed covered with blankets, a moment of great anxiety followed, which was, however, much relieved when, on stepping into the boat, he was accosted by Wishart, though in a feeble voice, and with an aspect pale as death from excessive bleeding. Directions having been immediately given to the coxswain to apply to Mr. Kennedy at the workyard to procure the best surgical aid, the boat was sent off without delay to Arbroath. The writer then landed at the rock, when the crane was in a very short time got into its place and again put in a working state.

Monday, 3rd July. The writer having come to Arbroath with the yacht, had an opportunity of visiting Michael Wishart, the artificer who had met with so severe an accident at the rock on the 30th ult., and had the pleasure to find him in a state of recovery. From Dr. Stevenson's account, under whose charge he had been placed, hopes were entertained that amputation would not be necessary, as his patient still kept free of fever or any appearance of mortification; and Wishart expressed a hope that he might, at least, be ultimately capable of keeping the light at the Bell Rock, as it was not now likely that he would assist further in building the house.

Saturday, 8th July.

It was remarked to-day, with no small demonstration of joy, that the tide, being neap, did not, for the first time, overflow the building at high-water. Flags were accordingly hoisted on the beacon-house, and crane on the top of the building, which were repeated from the floating light, Lighthouse yacht, tender, *Smeaton*, *Patriot*, and the two praams. A salute of three guns was also fired from the yacht at high-water, when, all the artificers being collected on the top of the building, three cheers were given in testimony of this important circumstance. A glass of rum was then served out to all hands on the rock and on board of the respective ships.

Sunday, 16th July.

Besides laying, boring, trenailing, wedging, and grouting thirty-two stones, several other operations were proceeded with on the rock at low-water, when some of the artificers were employed at the railways, and at high-water at the beacon-house. The seamen having prepared a quantity of tarpaulin, or cloth laid over with successive coats of hot tar, the joiners had just completed the covering of the roof with it. This sort of covering was lighter and more easily managed than sheet-lead in such a situation. As a further defence against the weather the whole exterior of this temporary residence was painted with three coats of white-lead paint. Between the timber framing of the habitable part of the beacon the interstices were to be stuffed with moss, as a light substance that would resist dampness and check sifting winds; the whole interior was then to be lined with green baize cloth, so that both without and within the cabins were to have a very comfortable appearance.

M

1809 Although the building artificers generally remained on the rock throughout the day, and the millwrights, joiners, and smiths, while their number was considerable, remained also during the night, yet the tender had hitherto been considered as their night quarters. But the wind having in the course of the day shifted to the N.W., and as the passage to the tender, in the boats, was likely to be attended with difficulty, the whole of the artificers, with Mr. Logan, the foreman, preferred remaining all night on the beacon, which had of late become the solitary abode of George Forsyth, a jobbing upholsterer, who had been employed in lining the beacon-house with cloth and in fitting up the bedding. Forsyth was a tall, thin, and rather loose-made man, who had an utter aversion at climbing upon the trap-ladders of the beacon, but especially at the process of boating, and the motion of the ship, which he said 'was death itself.' He therefore pertinaciously insisted with the landing-master in being left upon the beacon, with a small black dog as his only companion. The writer, however, felt some delicacy in leaving a single individual upon the rock, who must have been so very helpless in case of accident. This fabric had, from the beginning, been rather intended by the writer to guard against accident from the loss or damage of a boat, and as a place for making mortar, a smith's shop, and a store for tools during the working months, than as permanent quarters; nor was it at all meant to be possessed until the joiner-work was completely finished, and his own cabin, and that for the foreman, in readiness, when it was still to be left to the choice of the artificers to occupy the tender or the beacon. He, however, considered Forsyth's partiality and confidence in the latter as rather a fortunate occurrence.

1809 Wednesday, 19th July.

The whole of the artificers, twenty-three in number, now removed of their own accord from the tender, to lodge in the beacon, together with Peter Fortune, a person singularly adapted for a residence of this kind, both from the urbanity of his manners and the versatility of his talents. Fortune, in his person, was of small stature, and rather corpulent. Besides being a good Scots cook, he had acted both as groom and house-servant; he had been a soldier, a sutler, a writer's clerk, and an apothecary, from which he possessed the art of writing and suggesting recipes, and had hence, also, perhaps, acquired a turn for making collections in natural history. But in his practice in surgery on the Bell Rock, for which he received an annual fee of three guineas, he is supposed to have been rather partial to the use of the lancet. In short, Peter was the *factotum* of the beacon-house, where he ostensibly acted in the several capacities of cook, steward, surgeon, and barber, and kept a statement of the rations or expenditure of the provisions with the strictest integrity.

In the present important state of the building, when it had just attained the height of sixteen feet, and the upper courses, and especially the imperfect one, were in the wash of the heaviest seas, an express boat arrived at the rock with a letter from Mr. Kennedy, of the workyard, stating that in consequence of the intended expedition to Walcheren, an embargo had been laid on shipping at all the ports of Great Britain: that both the *Smeaton* and *Patriot* were detained at Arbroath, and that but for the proper view which Mr. Ramsey, the port officer, had taken of his orders, neither the express boat nor one which had been sent with provisions and necessaries for the floating light would have been permitted to leave the

1809 harbour. The writer set off without delay for Arbroath, and on landing used every possible means with the official people, but their orders were deemed so peremptory that even boats were not permitted to sail from any port upon the coast. In the meantime, the collector of the Customs at Montrose applied to the Board at Edinburgh, but could, of himself, grant no relief to the Bell Rock shipping.

At this critical period Mr. Adam Duff, then Sheriff of Forfarshire, now of the county of Edinburgh, and *ex officio* one of the Commissioners of the Northern Lighthouses, happened to be at Arbroath. Mr. Duff took an immediate interest in representing the circumstances of the case to the Board of Customs at Edinburgh. But such were the doubts entertained on the subject that, on having previously received the appeal from the collector at Montrose, the case had been submitted to the consideration of the Lords of the Treasury, whose decision was now waited for.

In this state of things the writer felt particularly desirous to get the thirteenth course finished, that the building might be in a more secure state in the event of bad weather. An opportunity was therefore embraced on the 25th, in sailing with provisions for the floating light, to carry the necessary stones to the rock for this purpose, which were landed and built on the 26th and 27th. But so closely was the watch kept up that a Custom-house officer was always placed on board of the *Smeaton* and *Patriot* while they were afloat, till the embargo was especially removed from the lighthouse vessels. The artificers at the Bell Rock had been reduced to fifteen, who were regularly supplied with provisions, along with the crew of the floating light, mainly through the port officer's liberal interpretation of his orders.

1809

Tuesday, 1st Aug.

There being a considerable swell and breach of sea upon the rock yesterday, the stones could not be got landed till the day following, when the wind shifted to the southward and the weather improved. But to-day no less than seventy-eight blocks of stone were landed, of which forty were built, which completed the fourteenth and part of the fifteenth courses. The number of workmen now resident in the beacon-house was augmented to twenty-four, including the landing-master's crew from the tender and the boat's crew from the floating light, who assisted at landing the stones. Those daily at work upon the rock at this period amounted to forty-six. A cabin had been laid out for the writer on the beacon, but his apartment had been the last which was finished, and he had not yet taken possession of it; for though he generally spent the greater part of the day, at this time, upon the rock, yet he always slept on board of the tender.

Friday, 11th Aug.

The wind was at S.E. on the 11th, and there was so very heavy a swell of sea upon the rock that no boat could approach it.

Saturday, 12th Aug.

The gale still continuing from the S.E., the sea broke with great violence both upon the building and the beacon. The former being twenty-three feet in height, the upper part of the crane erected on it having been lifted from course to course as the building advanced, was now about thirty-six feet above the rock. From observations made on the rise of the sea by this crane, the artificers were enabled to estimate its height to be about fifty feet above the rock, while the sprays fell with a most alarming noise upon their cabins. At low-water, in the evening, a signal was made from the beacon, at the earnest desire of some of the artificers, for the boats to come to the rock;

1809 and although this could not be effected without considerable hazard, it was, however, accomplished, when twelve of their number, being much afraid, applied to the foreman to be relieved, and went on board of the tender. But the remaining fourteen continued on the rock, with Mr. Peter Logan, the foreman builder. Although this rule of allowing an option to every man either to remain on the rock or return to the tender was strictly adhered to, yet, as it would have been extremely inconvenient to have the men parcelled out in this manner, it became necessary to embrace the first opportunity of sending those who had left the beacon to the workyard, with as little appearance of intention as possible, lest it should hurt their feelings, or prevent others from acting according to their wishes, either in landing on the rock or remaining on the beacon.

Tuesday, 15th Aug.

The wind had fortunately shifted to the S.W. this morning, and though a considerable breach was still upon the rock, yet the landing-master's crew were enabled to get one praam-boat, lightly loaded with five stones, brought in safety to the western creek; these stones were immediately laid by the artificers, who gladly embraced the return of good weather to proceed with their operations. The writer had this day taken possession of his cabin in the beacon-house. It was small, but commodious, and was found particularly convenient in coarse and blowing weather, instead of being obliged to make a passage to the tender in an open boat at all times, both during the day and the night, which was often attended with much difficulty and danger.

Saturday, 19th Aug.

For some days past the weather had been occasionally so thick and foggy that no small difficulty was experienced in going even between the rock and

1809

the tender, though quite at hand. But the floating light's boat lost her way so far in returning on board that the first land she made, after rowing all night, was Fifeness, a distance of about fourteen miles. The weather having cleared in the morning, the crew stood off again for the floating light, and got on board in a half-famished and much exhausted state, having been constantly rowing for about sixteen hours.

Sunday, 20th Aug.

The weather being very favourable to-day, fifty-three stones were landed, and the builders were not a little gratified in having built the twenty-second course, consisting of fifty-one stones, being the first course which had been completed in one day. This, as a matter of course, produced three hearty cheers. At twelve noon prayers were read for the first time on the Bell Rock; those present, counting thirty, were crowded into the upper apartment of the beacon, where the writer took a central position, while two of the artificers, joining hands, supported the Bible.

Friday, 25th Aug.

To-day the artificers laid forty-five stones, which completed the twenty-fourth course, reckoning above the first entire one, and the twenty-sixth above the rock. This finished the solid part of the building, and terminated the height of the outward casing of granite, which is thirty-one feet six inches above the rock or site of the foundation-stone, and about seventeen feet above high-water of spring-tides. Being a particular crisis in the progress of the lighthouse, the landing and laying of the last stone for the season was observed with the usual ceremonies.

From observations often made by the writer, in so far as such can be ascertained, it appears that no wave in the open seas, in an unbroken state, rises more than from seven to nine feet above the general surface of the ocean. The Bell Rock Lighthouse may therefore

1809 now be considered at from eight to ten feet above the height of the waves; and, although the sprays and heavy seas have often been observed, in the present state of the building, to rise to the height of fifty feet, and fall with a tremendous noise on the beacon-house, yet such seas were not likely to make any impression on a mass of solid masonry, containing about 1400 tons.

Wednesday, 30th Aug. The whole of the artificers left the rock at mid-day, when the tender made sail for Arbroath, which she reached about six p.m. The vessel being decorated with colours, and having fired a salute of three guns on approaching the harbour, the workyard artificers, with a multitude of people, assembled at the harbour, when mutual cheering and congratulations took place between those afloat and those on the quays. The tender had now, with little exception, been six months on the station at the Bell Rock, and during the last four months few of the squad of builders had been ashore. In particular, Mr. Peter Logan, the foreman, and Mr. Robert Selkirk, principal builder, had never once left the rock. The artificers, having made good wages during their stay, like seamen upon a return voyage, were extremely happy, and spent the evening with much innocent mirth and jollity.

In reflecting upon the state of the matters at the Bell Rock during the working months, when the writer was much with the artificers, nothing can equal the happy manner in which these excellent workmen spent their time. They always went from Arbroath to their arduous task cheering, and they generally returned in the same hearty state. While at the rock, between the tides, they amused themselves in reading, fishing, music, playing cards, draughts, etc., or in sporting with one another. In the workyard at Arbroath the young men were almost, without exception, employed in the evening at school, in writing and arithmetic, and not a few were learning architectural drawing, for which they had every convenience and facility, and were, in a very obliging manner, assisted in their studies by Mr. David Logan, clerk of the works. It therefore affords the most pleasing reflections to look back upon the pursuits of about sixty individuals who for years conducted themselves, on all occasions, in a sober and rational manner. 1809

IV

OPERATIONS OF 1810

1810 Thursday, 10th May

The wind had shifted to-day to W.N.W., when the writer, with considerable difficulty, was enabled to land upon the rock for the first time this season, at ten a.m. Upon examining the state of the building, and apparatus in general, he had the satisfaction to find everything in good order. The mortar in all the joints was perfectly entire. The building, now thirty feet in height, was thickly coated with *fuci* to the height of about fifteen feet, calculating from the rock: on the eastern side, indeed, the growth of seaweed was observable to the full height of thirty feet, and even on the top or upper bed of the last-laid course, especially towards the eastern side, it had germinated, so as to render walking upon it somewhat difficult.

The beacon-house was in a perfectly sound state, and apparently just as it had been left in the month of November. But the tides being neap, the lower parts, particularly where the beams rested on the rock, could not now be seen. The floor of the mortar gallery having been already laid down by Mr. Watt

1810 and his men on a former visit, was merely soaked with the sprays; but the joisting-beams which supported it had, in the course of the winter, been covered with a fine downy conferva produced by the range of the sea. They were also a good deal whitened with the mute of the cormorant and other sea-fowls, which had roosted upon the beacon in winter. Upon ascending to the apartments, it was found that the motion of the sea had thrown open the door of the cook-house: this was only shut with a single latch, that in case of shipwreck at the Bell Rock the mariner might find ready access to the shelter of this forlorn habitation, where a supply of provisions was kept; and being within two miles and a half of the floating light, a signal could readily be observed, when a boat might be sent to his relief as the weather permitted. An arrangement for this purpose formed one of the instructions on board of the floating light, but happily no instance occurred for putting it in practice. The hearth or fireplace of the cook-house was built of brick in as secure a manner as possible, to prevent accident from fire; but some of the plaster-work had shaken loose, from its damp state and the tremulous motion of the beacon in stormy weather. The writer next ascended to the floor which was occupied by the cabins of himself and his assistants, which were in tolerably good order, having only a damp and musty smell. The barrack for the artificers, over all, was next visited; it had now a very dreary and deserted appearance when its former thronged state was recollected. In some parts the water had come through the boarding, and had discoloured the lining of green cloth, but it was, nevertheless, in a good habitable condition. While the seamen were employed in landing a stock of provisions, a few of the artificers set to work with great eagerness to sweep and clean the several apartments. The exterior of the beacon was, in the meantime, examined, and found in perfect order. The painting, though it had a somewhat blanched appearance, adhered firmly both on the sides and roof, and only two or three panes of glass were broken in the cupola, which had either been blown out by the force of the wind, or perhaps broken by sea-fowl.

Having on this occasion continued upon the building and beacon a considerable time after the tide had begun to flow, the artificers were occupied in removing the forge from the top of the building, to which the gangway or wooden bridge gave great facility; and, although it stretched or had a span of forty-two feet, its construction was extremely simple, while the roadway was perfectly firm and steady. In returning from this visit to the rock every one was pretty well soused in spray before reaching the tender at two o'clock p.m., where things awaited the landing party in as comfortable a way as such a situation would admit.

Friday, 11th May.

The wind was still easterly, accompanied with rather a heavy swell of sea for the operations in hand. A landing was, however, made this morning, when the artificers were immediately employed in scraping the seaweed off the upper course of the building, in order to apply the moulds of the first course of the staircase, that the joggle-holes might be marked off in the upper course of the solid. This was also necessary previously to the writer's fixing the position of the entrance door, which was regulated chiefly by the appearance of the growth of the seaweed on the building, indicating the direction of the heaviest seas, on the opposite side of which the door was placed. The landing-master's crew succeeded in towing into the creek on the western side of the rock the praam-boat with the

1810 balance-crane, which had now been on board of the praam for five days. The several pieces of this machine, having been conveyed along the railways upon the waggons to a position immediately under the bridge, were elevated to its level, or thirty feet above the rock, in the following manner. A chain-tackle was suspended over a pulley from the cross-beam connecting the tops of the kingposts of the bridge, which was worked by a winch-machine with wheel, pinion, and barrel, round which last the chain was wound. This apparatus was placed on the beacon side of the bridge, at the distance of about twelve feet from the cross-beam and pulley in the middle of the bridge. Immediately under the cross-beam a hatch was formed in the roadway of the bridge, measuring seven feet in length and five feet in breadth, made to shut with folding boards like a double door, through which stones and other articles were raised; the folding doors were then let down, and the stone or load was gently lowered upon a waggon which was wheeled on railway trucks towards the lighthouse. In this manner the several castings of the balance-crane were got up to the top of the solid of the building.

The several apartments of the beacon-house having been cleaned out and supplied with bedding, a sufficient stock of provisions was put into the store, when Peter Fortune, formerly noticed, lighted his fire in the beacon for the first time this season. Sixteen artificers at the same time mounted to their barrack-room, and all the foremen of the works also took possession of their cabin, all heartily rejoiced at getting rid of the trouble of boating and the sickly motion of the tender.

Saturday, 12th May. The wind was at E.N.E., blowing so fresh, and accompanied with so much sea, that no stones could be landed to-day. The people on the rock, however, were busily employed in screwing together the balance-crane, cutting out the joggle-holes in the upper course, and preparing all things for commencing the building operations.

1810

The weather still continues boisterous, although the barometer has all the while stood at about 30 inches. Towards evening the wind blew so fresh at E. by S. that the boats both of the *Smeaton* and tender were obliged to be hoisted in, and it was feared that the *Smeaton* would have to slip her moorings. The people on the rock were seen busily employed, and had the balance-crane apparently ready for use, but no communication could be had with them to-day.

Sunday, 13th May.

The wind continued to blow so fresh, and the *Smeaton* rode so heavily with her cargo, that at noon a signal was made for her getting under weigh, when she stood towards Arbroath; and on board of the tender we are still without any communication with the people on the rock, where the sea was seen breaking over the top of the building in great sprays, and raging with much agitation among the beams of the beacon.

Monday, 14th May.

The wind, in the course of the day, had shifted from north to west; the sea being also considerably less, a boat landed on the rock at six p.m., for the first time since the 11th, with the provisions and water brought off by the *Patriot*. The inhabitants of the beacon were all well, but tired above measure for want of employment, as the balance-crane and apparatus was all in readiness. Under these circumstances they felt no less desirous of the return of good weather than those afloat, who were continually tossed with the agitation of the sea. The writer, in particular, felt himself almost as much fatigued and worn-out as he had been

Thursday, 17th May.

1810 at any period since the commencement of the work. The very backward state of the weather at so advanced a period of the season unavoidably created some alarm, lest he should be overtaken with bad weather at a late period of the season, with the building operations in an unfinished state. These apprehensions were, no doubt, rather increased by the inconveniences of his situation afloat, as the tender rolled and pitched excessively at times. This being also his first off-set for the season, every bone of his body felt sore with preserving a sitting posture while he endeavoured to pass away the time in reading; as for writing, it was wholly impracticable. He had several times entertained thoughts of leaving the station for a few days and going into Arbroath with the tender till the weather should improve; but as the artificers had been landed on the rock he was averse to this at the commencement of the season, knowing also that he would be equally uneasy in every situation till the first cargo was landed: and he therefore resolved to continue at his post until this should be effected.

Friday, 18th May. The wind being now N.W., the sea was considerably run down, and this morning at five o'clock the landing-master's crew, thirteen in number, left the tender; and having now no detention with the landing of artificers, they proceeded to unmoor the *Heddernwick* praam-boat, and towed her alongside of the *Smeaton*: and in the course of the day twenty-three blocks of stone, three casks of pozzolano, three of sand, three of lime, and one of Roman cement, together with three bundles of trenails and three of wedges, were all landed on the rock and raised to the top of the building by means of the tackle suspended from the cross-beam on the middle of the bridge. The stones were then moved along the bridge on the waggon to the building within reach of the balance-crane, with which they were laid in their respective places on the building. The masons immediately thereafter proceeded to bore the trenail-holes into the course below, and otherwise to complete the one in hand. When the first stone was to be suspended by the balance-crane, the bell on the beacon was rung, and all the artificers and seamen were collected on the building. Three hearty cheers were given while it was lowered into its place, and the steward served round a glass of rum, when success was drunk to the further progress of the building.

1810

Sunday, 20th May.

The wind was southerly to-day, but there was much less sea than yesterday, and the landing-master's crew were enabled to discharge and land twenty-three pieces of stone and other articles for the work. The artificers had completed the laying of the twenty-seventh or first course of the staircase this morning, and in the evening they finished the boring, trenailing, wedging, and grouting it with mortar. At twelve o'clock noon the beacon-house bell was rung, and all hands were collected on the top of the building, where prayers were read for the first time on the lighthouse, which forcibly struck every one, and had, upon the whole, a very impressive effect.

From the hazardous situation of the beacon-house with regard to fire, being composed wholly of timber, there was no small risk from accident: and on this account one of the most steady of the artificers was appointed to see that the fire of the cooking-house, and the lights in general, were carefully extinguished at stated hours.

Monday, 4th June.

This being the birthday of our much-revered Sovereign King George III, now in the fiftieth year of his reign, the shipping of the Lighthouse service

1810 were this morning decorated with colours according to the taste of their respective captains. Flags were also hoisted upon the beacon-house and balance-crane on the top of the building. At twelve noon a salute was fired from the tender, when the King's health was drunk, with all the honours, both on the rock and on board of the shipping.

Tuesday, 5th June. As the lighthouse advanced in height, the cubical contents of the stones were less, but they had to be raised to a greater height; and the walls, being thinner, were less commodious for the necessary machinery and the artificers employed, which considerably retarded the work. Inconvenience was also occasionally experienced from the men dropping their coats, hats, mallets, and other tools, at high-water, which were carried away by the tide; and the danger to the people themselves was now greatly increased. Had any of them fallen from the beacon or building at high-water, while the landing-master's crew were generally engaged with the craft at a distance, it must have rendered the accident doubly painful to those on the rock, who at this time had no boat, and consequently no means of rendering immediate and prompt assistance. In such cases it would have been too late to have got a boat by signal from the tender. A small boat, which could be lowered at pleasure, was therefore suspended by a pair of davits projected from the cook-house, the keel being about thirty feet from the rock. This boat, with its tackle, was put under the charge of James Glen, of whose exertions on the beacon mention has already been made, and who, having in early life been a seaman, was also very expert in the management of a boat. A life-buoy was likewise suspended from the bridge, to which a coil of line two hundred fathoms in length was attached, which could be let out to a person falling into the water, or to the people in the boat, should they not be able to work her with the oars. 1810

To-day twelve stones were landed on the rock, being the remainder of the *Patriot's* cargo; and the artificers built the thirty-ninth course, consisting of fourteen stones. The Bell Rock works had now a very busy appearance, as the lighthouse was daily getting more into form. Besides the artificers and their cook, the writer and his servant were also lodged on the beacon, counting in all twenty-nine; and at low-water the landing-master's crew, consisting of from twelve to fifteen seamen, were employed in transporting the building materials, working the landing apparatus on the rock, and dragging the stone waggons along the railways. Thursday, 7th June.

In the course of this day the weather varied much. In the morning it was calm, in the middle part of the day there were light airs of wind from the south, and in the evening fresh breezes from the east. The barometer in the writer's cabin in the beacon-house oscillated from 30 inches to 30·42, and the weather was extremely pleasant. This, in any situation, forms one of the chief comforts of life; but, as may easily be conceived, it was doubly so to people stuck, as it were, upon a pinnacle in the middle of the ocean. Friday, 8th June.

One of the praam-boats had been brought to the rock with eleven stones, notwithstanding the perplexity which attended the getting of those formerly landed taken up to the building. Mr. Peter Logan, the foreman builder, interposed, and prevented this cargo from being delivered; but the landing-master's crew were exceedingly averse to this arrangement, from an idea that "ill luck" would in future attend the praam, her cargo, and those who navigated her, from thus reversing her voyage. It may be noticed Sunday, 10th June.

1810 that this was the first instance of a praam-boat having been sent from the Bell Rock with any part of her cargo on board, and was considered so uncommon an occurrence that it became a topic of conversation among the seamen and artificers.

Tuesday, 12th June. To-day the stones formerly sent from the rock were safely landed, notwithstanding the augury of the seamen in consequence of their being sent away two days before.

Thursday, 14th June. To-day twenty-seven stones and eleven joggle-pieces were landed, part of which consisted of the forty-seventh course, forming the storeroom floor. The builders were at work this morning by four o'clock, in the hopes of being able to accomplish the laying of the eighteen stones of this course. But at eight o'clock in the evening they had still two to lay, and as the stones of this course were very unwieldy, being six feet in length, they required much precaution and care both in lifting and laying them. It was only on the writer's suggestion to Mr. Logan that the artificers were induced to leave off, as they had intended to complete this floor before going to bed. The two remaining stones were, however, laid in their places without mortar when the bell on the beacon was rung, and, all hands being collected on the top of the building, three hearty cheers were given on covering the first apartment. The steward then served out a dram to each, when the whole retired to their barrack much fatigued, but with the anticipation of the most perfect repose even in the "hurricane-house," amidst the dashing seas on the Bell Rock.

While the workmen were at breakfast and dinner it was the writer's usual practice to spend his time on the walls of the building, which, notwithstanding the narrowness of the track, nevertheless formed his principal walk when the rock was under water, But this afternoon he had his writing-desk set upon the storeroom floor, when he wrote to Mrs. Stevenson—certainly the first letter dated from the Bell Rock *Lighthouse*—giving a detail of the fortunate progress of the work, with an assurance that the lighthouse would soon be completed at the rate at which it now proceeded; and, the *Patriot* having sailed for Arbroath in the evening, he felt no small degree of pleasure in despatching this communication to his family.

1810

The weather still continuing favourable for the operations at the rock, the work proceeded with much energy, through the exertions both of the seamen and artificers. For the more speedy and effectual working of the several tackles in raising the materials as the building advanced in height, and there being a great extent of railway to attend to, which required constant repairs, two additional millwrights were added to the complement on the rock, which, including the writer, now counted thirty-one in all. So crowded was the men's barrack that the beds were ranged five tier in height, allowing only about one foot eight inches for each bed. The artificers commenced this morning at five o'clock, and, in the course of the day, they laid the forty-eighth and forty-ninth courses, consisting each of sixteen blocks. From the favourable state of the weather, and the regular manner in which the work now proceeded, the artificers had generally from four to seven extra hours' work, which, including their stated wages of 3s. 4d., yielded them from 5s. 4d. to about 6s. 10d. per day besides their board; even the postage of their letters was paid while they were at the Bell Rock. In these advantages the foremen also shared, having about double the pay and amount of premiums of the artificers. The seamen being less out

1810 of their element in the Bell Rock operations than the landsmen, their premiums consisted in a slump sum payable at the end of the season, which extended from three to ten guineas.

As the laying of the floors was somewhat tedious, the landing-master and his crew had got considerably beforehand with the building artificers in bringing materials faster to the rock than they could be built. The seamen having, therefore, some spare time, were occasionally employed during fine weather in dredging or grappling for the several mushroom anchors and mooring-chains which had been lost in the vicinity of the Bell Rock during the progress of the work by the breaking loose and drifting of the floating buoys. To encourage their exertions in this search, five guineas were offered as a premium for each set they should find; and, after much patient application, they succeeded to-day in hooking one of these lost anchors with its chain.

It was a general remark at the Bell Rock, as before noticed, that fish were never plenty in its neighbourhood excepting in good weather. Indeed, the seamen used to speculate about the state of the weather from their success in fishing. When the fish disappeared at the rock, it was considered a sure indication that a gale was not far off, as the fish seemed to seek shelter in deeper water from the roughness of the sea during these changes in the weather. At this time the rock, at high-water, was completely covered with podlies, or the fry of the coal-fish, about six or eight inches in length. The artificers sometimes occupied half an hour after breakfast and dinner in catching these little fishes, but were more frequently supplied from the boats of the tender.

Saturday, 16th June. The landing-master having this day discharged the

Smeaton and loaded the *Hedderwick* and *Dickie* praam-boats with nineteen stones, they were towed to their respective moorings, when Captain Wilson, in consequence of the heavy swell of sea, came in his boat to the beacon-house to consult with the writer as to the propriety of venturing the loaded praam-boats with their cargoes to the rock while so much sea was running. After some dubiety expressed on the subject, in which the ardent mind of the landing-master suggested many arguments in favour of his being able to convey the praams in perfect safety, it was acceded to. In bad weather, and especially on occasions of difficulty like the present, Mr. Wilson, who was an extremely active seaman, measuring about five feet three inches in height, of a robust habit, generally dressed himself in what he called a *monkey jacket,* made of thick duffle cloth, with a pair of Dutchman's petticoat trousers, reaching only to his knees, where they were met with a pair of long water-tight boots; with this dress, his glazed hat, and his small brass speaking-trumpet in his hand, he bade defiance to the weather. When he made his appearance in this most suitable attire for the service his crew seemed to possess additional life, never failing to use their utmost exertions when the captain put on his *storm rigging.* They had this morning commenced loading the praam-boats at four o'clock, and proceeded to tow them into the eastern landing-place, which was accomplished with much dexterity, though not without the risk of being thrown, by the force of the sea, on certain projecting ledges of the rock. In such a case the loss even of a single stone would have greatly retarded the work. For the greater safety in entering the creek it was necessary to put out several warps and guy-ropes to guide the boats into its narrow and intricate

1810 entrance; and it frequently happened that the sea made a clean breach over the praams, which not only washed their decks, but completely drenched the crew in water.

Sunday, 17th June. It was fortunate, in the present state of the weather, that the fiftieth course was in a sheltered spot, within the reach of the tackle of the winch-machine upon the bridge; a few stones were stowed upon the bridge itself, and the remainder upon the building, which kept the artificers at work. The stowing of the materials upon the rock was the department of Alexander Brebner, mason, who spared no pains in attending to the safety of the stones, and who, in the present state of the work, when the stones were landed faster than could be built, generally worked till the water rose to his middle. At one o'clock to-day the bell rung for prayers, and all hands were collected into the upper barrack-room of the beacon-house, when the usual service was performed.

The wind blew very hard in the course of last night from N.E., and to-day the sea ran so high that no boat could approach the rock. During the dinner-hour, when the writer was going to the top of the building as usual, but just as he had entered the door and was about to ascend the ladder, a great noise was heard overhead, and in an instant he was soused in water from a sea which had most unexpectedly come over the walls, though now about fifty-eight feet in height. On making his retreat he found himself completely whitened by the lime, which had mixed with the water while dashing down through the different floors; and, as nearly as he could guess, a quantity equal to about a hogshead had come over the walls, and now streamed out at the door. After having shifted himself, he again sat down in his cabin, the sea continuing

1810

to run so high that the builders did not resume their operations on the walls this afternoon. The incident just noticed did not create more surprise in the mind of the writer than the sublime appearance of the waves as they rolled majestically over the rock. This scene he greatly enjoyed while sitting at his cabin window; each wave approached the beacon like a vast scroll unfolding; and in passing discharged a quantity of air, which he not only distinctly felt, but was even sufficient to lift the leaves of a book which lay before him. These waves might be ten or twelve feet in height, and about 250 feet in length, their smaller end being towards the north, where the water was deep, and they were opened or cut through by the interposition of the building and beacon. The gradual manner in which the sea, upon these occasions, is observed to become calm or to subside, is a very remarkable feature of this phenomenon. For example, when a gale is succeeded by a calm, every third or fourth wave forms one of these great seas, which occur in spaces of from three to five minutes, as noted by the writer's watch; but in the course of the next tide they become less frequent, and take off so as to occur only in ten or fifteen minutes; and, singular enough, at the third tide after such gales, the writer has remarked that only one or two of these great waves appear in the course of the whole tide.

Tuesday, 19th June.

The 19th was a very unpleasant and disagreeable day, both for the seamen and artificers, as it rained throughout with little intermission from four a.m. till eleven p.m., accompanied with thunder and lightning, during which period the work nevertheless continued unremittingly, and the builders laid the fifty-first and fifty-second courses. This state of weather was no less severe upon the mortar-makers, who required to temper

1810 or prepare the mortar of a thicker or thinner consistency, in some measure, according to the state of the weather. From the elevated position of the building, the mortar gallery on the beacon was now much lower, and the lime-buckets were made to traverse upon a rope distended between it and the building. On occasions like the present, however, there was often a difference of opinion between the builders and the mortar-makers. John Watt, who had the principal charge of the mortar, was a most active worker, but, being somewhat of an irascible temper, the builders occasionally amused themselves at his expense; for while he was eagerly at work with his large iron-shod pestle in the mortar-tub, they often sent down contradictory orders, some crying, 'Make it a little stiffer, or thicker, John,' while others called out to make it 'thinner,' to which he generally returned very speedy and sharp replies, so that these conversations at times were rather amusing.

During wet weather the situation of the artificers on the top of the building was extremely disagreeable; for although their work did not require great exertion, yet, as each man had his particular part to perform, either in working the crane or in laying the stones, it required the closest application and attention, not only on the part of Mr. Peter Logan, the foreman, who was constantly on the walls, but also of the chief workmen. Robert Selkirk, the principal builder, for example, had every stone to lay in its place. David Cumming, a mason, had the charge of working the tackle of the balance-weight, and James Scott, also a mason, took charge of the purchase with which the stones were laid; while the pointing the joints of the walls with cement was intrusted to William Reid and William Kennedy, who stood upon a scaffold suspended over the walls in rather a frightful manner. The least act of carelessness or inattention on the part of any of these men might have been fatal, not only to themselves, but also to the surrounding workmen, especially if any accident had happened to the crane itself, while the material damage or loss of a single stone would have put an entire stop to the operations until another could have been brought from Arbroath. The artificers, having wrought seven and a half hours of extra time to-day, had 3s. 9d. of extra pay, while the foremen had 7s. 6d. over and above their stated pay and board. Although, therefore, the work was both hazardous and fatiguing, yet, the encouragement being considerable, they were always very cheerful, and perfectly reconciled to the confinement and other disadvantages of the place.

During fine weather, and while the nights were short, the duty on board of the floating light was literally nothing but a waiting on, and therefore one of her boats, with a crew of five men, daily attended the rock, but always returned to the vessel at night. The carpenter, however, was one of those who was left on board of the ship, as he also acted in the capacity of assistant lightkeeper, being, besides, a person who was apt to feel discontent and to be averse to changing his quarters, especially to work with the millwrights and joiners at the rock, who often, for hours together, wrought knee-deep, and not unfrequently up to the middle, in water. Mr. Watt having about this time made a requisition for another hand, the carpenter was ordered to attend the rock in the floating light's boat. This he did with great reluctance, and found so much fault that he soon got into discredit with his messmates. On this occasion he left the Lighthouse service, and went as a sailor

1810

in a vessel bound for America—a step which, it is believed, he soon regretted, as, in the course of things, he would, in all probability, have accompanied Mr. John Reid, the principal lightkeeper of the floating light, to the Bell Rock Lighthouse as his principal assistant. The writer had a wish to be of service to this man, as he was one of those who came off to the floating light in the month of September 1807, while she was riding at single anchor after the severe gale of the 7th, at a time when it was hardly possible to make up this vessel's crew; but the crossness of his manner prevented his reaping the benefit of such intentions.

Friday, 22nd June.

The building operations had for some time proceeded more slowly, from the higher parts of the lighthouse requiring much longer time than an equal tonnage of the lower courses. The duty of the landing-master's crew had, upon the whole, been easy of late; for though the work was occasionally irregular, yet the stones being lighter, they were more speedily lifted from the hold of the stone vessel to the deck of the praam-boat, and again to the waggons on the railway, after which they came properly under the charge of the foreman builder. It is, however, a strange, though not an uncommon, feature in the human character, that, when people have least to complain of, they are most apt to become dissatisfied, as was now the case with the seamen employed in the Bell Rock service about their rations of beer. Indeed, ever since the carpenter of the floating light, formerly noticed, had been brought to the rock, expressions of discontent had been manifested upon various occasions. This being represented to the writer, he sent for Captain Wilson, the landing-master, and Mr. Taylor, commander of the tender, with whom he talked over the subject. They stated that they considered the daily allowance of the seamen in every respect ample, and that, the work being now much lighter than formerly, they had no just ground for complaint; Mr. Taylor adding that, if those who now complained 'were even to be fed upon soft bread and turkeys, they would not think themselves right.' At twelve noon the work of the landing-master's crew was completed for the day; but at four o'clock, while the rock was under water, those on the beacon were surprised by the arrival of a boat from the tender without any signal having been made from the beacon. It brought the following note to the writer from the landing-master's crew:—

'*Sir Joseph Banks Tender.*

'Sir,—We are informed by our masters that our allowance is to be as before, and it is not sufficient to serve us, for we have been at work since four o'clock this morning, and we have come on board to dinner, and there is no beer for us before to-morrow morning, to which a sufficient answer is required before we go from the beacon; and we are, Sir, your most obedient servants.'

On reading this, the writer returned a verbal message, intimating that an answer would be sent on board of the tender, at the same time ordering the boat instantly to quit the beacon. He then addressed the following note to the landing-master:—

'*Beacon-house, 22nd June* 1810,
Five o'clock p.m.

'Sir,—I have just now received a letter purporting to be from the landing-master's crew and seamen on

1810 board of the *Sir Joseph Banks,* though without either date or signature; in answer to which I enclose a statement of the daily allowance of provisions for the seamen in this service, which you will post up in the ship's galley, and at seven o'clock this evening I will come on board to inquire into this unexpected and most unnecessary demand for an additional allowance of beer. In the enclosed you will not find any alteration from the original statement, fixed in the galley at the beginning of the season. I have, however, judged this mode of giving your people an answer preferable to that of conversing with them on the beacon.—I am, Sir, your most obedient servant, ROBERT STEVENSON.

'TO CAPTAIN WILSON.'

'*Beacon House, 22nd June* 1810.—Schedule of the daily allowance of provisions to be served out on board of the *Sir Joseph Banks* tender: "1½ lb. beef; 1 lb. bread; 8 oz. oatmeal; 2 oz. barley; 2 oz. butter; 3 quarts beer; vegetables and salt no stated allowance. When the seamen are employed in unloading the *Smeaton* and *Patriot,* a draught of beer is, as formerly, to be allowed from the stock of these vessels. Further, in wet and stormy weather, or when the work commences very early in the morning, or continues till a late hour at night, a glass of spirits will also be served out to the crew as heretofore, on the requisition of the landing-master." ROBERT STEVENSON.'

On writing this letter and schedule, a signal was made on the beacon for the landing-master's boat, which immediately came to the rock, and the schedule was afterwards stuck up in the tender's galley. When sufficient time had been allowed to the crew to consider of their conduct, a second signal was made for a boat, and at seven o'clock the writer left the Bell Rock, after a residence of four successive weeks in the beacon-house. The first thing which occupied his attention on board of the tender was to look round upon the lighthouse, which he saw, with some degree of emotion and surprise, now vying in height with the beacon-house; for although he had often viewed it from the extremity of the western railway on the rock, yet the scene, upon the whole, seemed far more interesting from the tender's moorings at the distance of about half a mile.

The *Smeaton* having just arrived at her moorings with a cargo, a signal was made for Captain Pool to come on board of the tender, that he might be at hand to remove from the service any of those who might persist in their discontented conduct. One of the two principal leaders in this affair, the master of one of the praam-boats, who had also steered the boat which brought the letter to the beacon, was first called upon deck, and asked if he had read the statement fixed up in the galley this afternoon, and whether he was satisfied with it. He replied that he had read the paper, but was not satisfied, as it held out no alteration in the allowance, on which he was immediately ordered into the *Smeaton's* boat. The next man called had but lately entered the service, and, being also interrogated as to his resolution, he declared himself to be of the same mind with the praam-master, and was also forthwith ordered into the boat. The writer, without calling any more of the seamen, went forward to the gangway, where they were collected and listening to what was passing upon deck. He addressed them at the hatchway, and stated that two of their companions had just been dismissed the service and sent on board

1810 of the *Smeaton* to be conveyed to Arbroath. He therefore wished each man to consider for himself how far it would be proper, by any unreasonableness of conduct, to place themselves in a similar situation, especially as they were aware that it was optional in him either to dismiss them or send them on board a man-of-war. It might appear that much inconveniency would be felt at the rock by a change of hands at this critical period, by checking for a time the progress of a building so intimately connected with the best interests of navigation; yet this would be but of a temporary nature, while the injury to themselves might be irreparable. It was now, therefore, required of any man who, in this disgraceful manner, chose to leave the service, that he should instantly make his appearance on deck while the *Smeaton's* boat was alongside. But those below having expressed themselves satisfied with their situation—viz., William Brown, George Gibb, Alexander Scott, John Dick, Robert Couper, Alexander Shephard, James Grieve, David Carey, William Pearson, Stuart Eaton, Alexander Lawrence, and John Spink—were accordingly considered as having returned to their duty. This disposition to mutiny, which had so strongly manifested itself, being now happily suppressed, Captain Pool got orders to proceed for Arbroath Bay, and land the two men he had on board, and to deliver the following letter at the office of the workyard:—

'*On board of the Tender off the Bell Rock,*
22nd June 1810, *eight o'clock p.m.*

'Dear Sir,—A discontented and mutinous spirit having manifested itself of late among the landing-master's crew, they struck work to-day and demanded

1810

an additional allowance of beer, and I have found it necessary to dismiss D——d and M——e, who are now sent on shore with the *Smeaton*. You will therefore be so good as to pay them their wages, including this day only. Nothing can be more unreasonable than the conduct of the seamen on this occasion, as the landing-master's crew not only had their allowance on board of the tender, but, in the course of this day, they had drawn no fewer than twenty-four quart pots of beer from the stock of the *Patriot* while unloading her.—I remain, yours truly, ROBERT STEVENSON.

'To Mr. LACHLAN KENNEDY,
Bell Rock Office, Arbroath.'

On despatching this letter to Mr. Kennedy, the writer returned to the beacon about nine o'clock, where this afternoon's business had produced many conjectures, especially when the *Smeaton* got under weigh, instead of proceeding to land her cargo. The bell on the beacon being rung, the artificers were assembled on the bridge, when the affair was explained to them. He, at the same time, congratulated them upon the first appearance of mutiny being happily set at rest by the dismissal of its two principal abettors.

Sunday, 24th June.

At the rock the landing of the materials and the building operations of the light-room store went on successfully, and in a way similar to those of the provision store. To-day it blew fresh breezes; but the seamen nevertheless landed twenty-eight stones, and the artificers built the fifty-eighth and fifty-ninth courses. The works were visited by Mr. Murdoch, junior, from Messrs. Boulton and Watt's works of Soho. He landed just as the bell rung for prayers, after which the writer enjoyed much pleasure from

1810 his very intelligent conversation; and, having been almost the only stranger he had seen for some weeks, he parted with him, after a short interview, with much regret.

Thursday, 28th June. Last night the wind had shifted to north-east, and, blowing fresh, was accompanied with a heavy surf upon the rock. Towards high-water it had a very grand and wonderful appearance. Waves of considerable magnitude rose as high as the solid or level of the entrance-door, which, being open to the south-west, was fortunately to the leeward; but on the windward side the sprays flew like lightning up the sloping sides of the building; and although the walls were now elevated sixty-four feet above the rock, and about fifty-two feet from high-water mark, yet the artificers were nevertheless wetted, and occasionally interrupted, in their operations on the top of the walls. These appearances were, in a great measure, new at the Bell Rock, there having till of late been no building to conduct the seas, or object to compare with them. Although, from the description of the Eddystone Lighthouse, the mind was prepared for such effects, yet they were not expected to the present extent in the summer season; the sea being most awful to-day, whether observed from the beacon or the building. To windward, the sprays fell from the height above noticed in the most wonderful cascades, and streamed down the walls of the building in froth as white as snow. To leeward of the lighthouse the collision or meeting of the waves produced a pure white kind of *drift*; it rose about thirty feet in height, like a fine downy mist, which, in its fall, fell upon the face and hands more like a dry powder than a liquid substance. The effect of these seas, as they raged among the beams and dashed upon the higher parts of the beacon, produced a temporary tremulous motion throughout the whole fabric, which to a stranger must have been frightful.

1810

Sunday, 1st July.

The writer had now been at the Bell Rock since the latter end of May, or about six weeks, during four of which he had been a constant inhabitant of the beacon without having been once off the rock. After witnessing the laying of the sixty-seventh or second course of the bedroom apartment, he left the rock with the tender and went ashore, as some arrangements were to be made for the future conduct of the works at Arbroath, which were soon to be brought to a close; the landing-master's crew having, in the meantime, shifted on board of the *Patriot*. In leaving the rock, the writer kept his eyes fixed upon the lighthouse, which had recently got into the form of a house, having several tiers or stories of windows. Nor was he unmindful of his habitation in the beacon —now far overtopped by the masonry,—where he had spent several weeks in a kind of active retirement, making practical experiment of the fewness of the positive wants of man. His cabin measured not more than four feet three inches in breadth on the floor; and though, from the oblique direction of the beams of the beacon, it widened towards the top, yet it did not admit of the full extension of his arms when he stood on the floor; while its length was little more than sufficient for suspending a cot-bed during the night, calculated for being triced up to the roof through the day, which left free room for the admission of occasional visitants. His folding table was attached with hinges, immediately under the small window of the apartment, and his books, barometer, thermometer, portmanteau, and two or three camp-stools, formed the bulk of his movables.

1810

His diet being plain, the paraphernalia of the table were proportionally simple; though everything had the appearance of comfort, and even of neatness, the walls being covered with green cloth formed into panels with red tape, and his bed festooned with curtains of yellow cotton-stuff. If, in speculating upon the abstract wants of man in such a state of exclusion, one were reduced to a single book, the Sacred Volume—whether considered for the striking diversity of its story, the morality of its doctrine, or the important truths of its gospel—would have proved by far the greatest treasure.

Monday, 2nd July.

In walking over the workyard at Arbroath this morning, the writer found that the stones of the course immediately under the cornice were all in hand, and that a week's work would now finish the whole, while the intermediate courses lay ready numbered and marked for shipping to the rock. Among other subjects which had occupied his attention to-day was a visit from some of the relations of George Dall, a young man who had been impressed near Dundee in the month of February last; a dispute had arisen between the magistrates of that burgh and the Regulating Officer as to his right of impressing Dall, who was *bonâ fide* one of the protected seamen in the Bell Rock service. In the meantime, the poor lad was detained, and ultimately committed to the prison of Dundee, to remain until the question should be tried before the Court of Session. His friends were naturally very desirous to have him relieved upon bail. But, as this was only to be done by the judgment of the Court, all that could be said was that his pay and allowances should be continued in the same manner as if he had been upon the sick-list. The circumstances of Dall's case were briefly these:—

He had gone to see some of his friends in the neighbourhood of Dundee, in winter, while the works were suspended, having got leave of absence from Mr. Taylor, who commanded the Bell Rock tender, and had in his possession one of the Protection Medals. Unfortunately, however, for Dall, the Regulating Officer thought proper to disregard these documents, as, according to the strict and literal interpretation of the Admiralty regulations, a seaman does not stand protected unless he is actually on board of his ship, or in a boat belonging to her, or has the Admiralty protection in his possession. This order of the Board, however, cannot be rigidly followed in practice; and therefore, when the matter is satisfactorily stated to the Regulating Officer, the impressed man is generally liberated. But in Dall's case this was peremptorily refused, and he was retained at the instance of the magistrates. The writer having brought the matter under the consideration of the Commissioners of the Northern Lighthouses, they authorised it to be tried on the part of the Lighthouse Board, as one of extreme hardship. The Court, upon the first hearing, ordered Dall to be liberated from prison; and the proceedings never went further.

1810

Being now within twelve courses of being ready for building the cornice, measures were taken for getting the stones of it and the parapet-wall of the light-room brought from Edinburgh, where, as before noticed, they had been prepared and were in readiness for shipping. The honour of conveying the upper part of the lighthouse, and of landing the last stone of the building on the rock, was considered to belong to Captain Pool of the *Smeaton*, who had been longer in the service than the master of the *Patriot*. The *Smeaton* was, therefore, now partly loaded with old

Wednesday, 4th July.

1810 iron, consisting of broken railways and other lumber which had been lying about the rock. After landing these at Arbroath, she took on board James Craw, with his horse and cart, which could now be spared at the workyard, to be employed in carting the stones from Edinburgh to Leith. Alexander Davidson and William Kennedy, two careful masons, were also sent to take charge of the loading of the stones at Greenside, and stowing them on board of the vessel at Leith. The writer also went on board, with a view to call at the Bell Rock and to take his passage up the Firth of Forth. The wind, however, coming to blow very fresh from the eastward, with thick and foggy weather, it became necessary to reef the mainsail and set the second jib. When in the act of making a tack towards the tender, the sailors who worked the head-sheets were, all of a sudden, alarmed with the sound of the smith's hammer and anvil on the beacon, and had just time to put the ship about to save her from running ashore on the north-western point of the rock, marked 'James Craw's Horse.' On looking towards the direction from whence the sound came, the building and beacon-house were seen, with consternation, while the ship was hailed by those on the rock, who were no less confounded at seeing the near approach of the *Smeaton*; and, just as the vessel cleared the danger, the smith and those in the mortar gallery made signs in token of their happiness at our fortunate escape. From this occurrence the writer had an experimental proof of the utility of the large bells which were in preparation to be rung by the machinery of the revolving light; for, had it not been the sound of the smith's anvil, the *Smeaton*, in all probability, would have been wrecked upon the rock. In case the vessel had struck, those on board might have been safe, having now the beacon-house as a place of refuge; but the vessel, which was going at a great velocity, must have suffered severely, and it was more than probable that the horse would have been drowned, there being no means of getting him out of the vessel. Of this valuable animal and his master we shall take an opportunity of saying more in another place.

1810

The weather cleared up in the course of the night, but the wind shifted to the N.E. and blew very fresh. From the force of the wind, being now the period of spring-tides, a very heavy swell was experienced at the rock. At two o'clock on the following morning the people on the beacon were in a state of great alarm about their safety, as the sea had broke up part of the floor of the mortar gallery, which was thus cleared of the lime-casks and other buoyant articles; and, the alarm-bell being rung, all hands were called to render what assistance was in their power for the safety of themselves and the materials. At this time some would willingly have left the beacon and gone into the building: the sea, however, ran so high that there was no passage along the bridge of communication, and, when the interior of the lighthouse came to be examined in the morning, it appeared that great quantities of water had come over the walls—now eighty feet in height—and had run down through the several apartments and out at the entrance door.

Thursday, 5th July.

The upper course of the lighthouse at the workyard of Arbroath was completed on the 6th, and the whole of the stones were, therefore, now ready for being shipped to the rock. From the present state of the works it was impossible that the two squads of artificers at Arbroath and the Bell Rock could meet together at this period; and as in public works of this

1810 kind, which had continued for a series of years, it is not customary to allow the men to separate without what is termed a "finishing-pint," five guineas were for this purpose placed at the disposal of Mr. David Logan, clerk of works. With this sum the stone-cutters at Arbroath had a merry meeting in their barrack, collected their sweethearts and friends, and concluded their labours with a dance. It was remarked, however, that their happiness on this occasion was not without alloy. The consideration of parting and leaving a steady and regular employment, to go in quest of work and mix with other society, after having been harmoniously lodged for years together in one large "guildhall or barrack," was rather painful.

Friday, 6th July.

While the writer was at Edinburgh he was fortunate enough to meet with Mrs. Dickson, only daughter of the late celebrated Mr. Smeaton, whose works at the Eddystone Lighthouse had been of such essential consequence to the operations at the Bell Rock. Even her own elegant accomplishments are identified with her father's work, she having herself made the drawing of the vignette on the title-page of the *Narrative of the Eddystone Lighthouse.* Every admirer of the works of that singularly eminent man must also feel an obligation to her for the very comprehensive and distinct account given of his life, which is attached to his reports, published, in three volumes quarto, by the Society of Civil Engineers. Mrs. Dickson, being at this time returning from a tour to the Hebrides and Western Highlands of Scotland, had heard of the Bell Rock works, and from their similarity to those of the Eddystone was strongly impressed with a desire of visiting the spot. But on inquiring for the writer at Edinburgh, and finding from him that the upper part of the lighthouse, consisting of nine courses, might be seen in the immediate vicinity, and also that one of the vessels, which, in compliment to her father's memory, had been named the *Smeaton*, might also now be seen in Leith, she considered herself extremely fortunate; and having first visited the works at Greenside, she afterwards went to Leith to see the *Smeaton*, then loading for the Bell Rock. On stepping on board, Mrs. Dickson seemed to be quite overcome with so many concurrent circumstances, tending in a peculiar manner to revive and enliven the memory of her departed father, and, on leaving the vessel, she would not be restrained from presenting the crew with a piece of money. The *Smeaton* had been named spontaneously, from a sense of the obligation which a public work of the description of the Bell Rock owed to the labours and abilities of Mr. Smeaton. The writer certainly never could have anticipated the satisfaction which he this day felt in witnessing the pleasure it afforded to the only representative of this great man's family.

1810

The gale from the N.E. still continued so strong, accompanied with a heavy sea, that the *Patriot* could not approach her moorings; and although the tender still kept her station, no landing was made to-day at the rock. At high-water it was remarked that the spray rose to the height of about sixty feet upon the building. The *Smeaton* now lay in Leith loaded, but, the wind and weather being so unfavourable for her getting down the Firth, she did not sail till this afternoon. It may be here proper to notice that the loading of the centre of the light-room floor, or last principal stone of the building, did not fail, when put on board, to excite an interest among those connected with the work. When the stone was laid upon the cart to be conveyed to Leith, the seamen fixed an

Friday, 20th July.

1810

ensign-staff and flag into the circular hole in the centre of the stone, and decorated their own hats, and that of James Craw, the Bell Rock carter, with ribbons; even his faithful and trusty horse Brassey was ornamented with bows and streamers of various colours. The masons also provided themselves with new aprons, and in this manner the cart was attended in its progress to the ship. When the cart came opposite the Trinity House of Leith, the officer of that corporation made his appearance dressed in his uniform, with his staff of office; and when it reached the harbour, the shipping in the different tiers where the *Smeaton* lay hoisted their colours, manifesting by these trifling ceremonies the interest with which the progress of this work was regarded by the public, as ultimately tending to afford safety and protection to the mariner. The wind had fortunately shifted to the S.W., and about five o'clock this afternoon the *Smeaton* reached the Bell Rock.

Friday, 27th July.

The artificers had finished the laying of the balcony course, excepting the centre-stone of the light-room floor, which, like the centres of the other floors, could not be laid in its place till after the removal of the foot and shaft of the balance-crane. During the dinner-hour, when the men were off work, the writer generally took some exercise by walking round the walls when the rock was under water; but to-day his boundary was greatly enlarged, for, instead of the narrow wall as a path, he felt no small degree of pleasure in walking round the balcony and passing out and in at the space allotted for the light-room door. In the labours of this day both the artificers and seamen felt their work to be extremely easy compared with what it had been for some days past.

Sunday, 29th July.

Captain Wilson and his crew had made preparations for landing the last stone, and, as may well be supposed, this was a day of great interest at the Bell Rock. 'That it might lose none of its honours,' as he expressed himself, the *Hedderwick* praam-boat, with which the first stone of the building had been landed, was appointed also to carry the last. At seven o'clock this evening the seamen hoisted three flags upon the *Hedderwick*, when the colours of the *Dickie* praam-boat, tender, *Smeaton*, floating light, beacon-house, and light-house were also displayed; and, the weather being remarkably fine, the whole presented a very gay appearance, and, in connection with the associations excited, the effect was very pleasing. The praam which carried the stone was towed by the seamen in gallant style to the rock, and, on its arrival, cheers were given as a finale to the landing department.

1810

Monday, 30th July.

The ninetieth or last course of the building having been laid to-day, which brought the masonry to the height of one hundred and two feet six inches, the lintel of the light-room door, being the finishing-stone of the exterior walls, was laid with due formality by the writer, who, at the same time, pronounced the following benediction: "May the Great Architect of the Universe, under whose blessing this perilous work has prospered, preserve it as a guide to the mariner."

Friday, 3rd Aug.

At three p.m., the necessary preparations having been made, the artificers commenced the completing of the floors of the several apartments, and at seven o'clock the centre-stone of the light-room floor was laid, which may be held as finishing the masonry of this important national edifice. After going through the usual ceremonies observed by the brotherhood on occasions of this kind, the writer, addressing himself to the artificers and seamen who were present, briefly alluded to the utility of the undertaking as a monument of the wealth of British commerce, erected

1810 through the spirited measures of the Commissioners of the Northern Lighthouses by means of the able assistance of those who now surrounded him. He then took an opportunity of stating that toward those connected with this arduous work he would ever retain the most heartfelt regard in all their interests.

Saturday, 4th Aug. When the bell was rung as usual on the beacon this morning, every one seemed as if he were at a loss what to make of himself. At this period the artificers at the rock consisted of eighteen masons, two joiners, one millwright, one smith, and one mortar-maker, besides Messrs. Peter Logan and Francis Watt, foremen, counting in all twenty-five; and matters were arranged for proceeding to Arbroath this afternoon with all hands. The *Sir Joseph Banks* tender had by this time been afloat, with little intermission, for six months, during greater part of which the artificers had been almost constantly off at the rock, and were now much in want of necessaries of almost every description. Not a few had lost different articles of clothing, which had dropped into the sea from the beacon and building. Some wanted jackets; others, from want of hats, wore nightcaps; each was, in fact, more or less curtailed in his wardrobe, and it must be confessed that at best the party were but in a very tattered condition. This morning was occupied in removing the artificers and their bedding on board of the tender; and although their personal luggage was easily shifted, the boats had, nevertheless, many articles to remove from the beacon-house, and were consequently employed in this service till eleven a.m. All hands being collected and just ready to embark, as the water had nearly overflowed the rock, the writer, in taking leave, after alluding to the harmony which had ever marked the conduct of those employed on the Bell Rock, took occasion to compliment the great zeal, attention, and abilities of Mr. Peter Logan and Mr. Francis Watt, foremen; Captain James Wilson, landing-master; and Captain David Taylor, commander of the tender, who, in their several departments, had so faithfully discharged the duties assigned to them, often under circumstances the most difficult and trying. The health of these gentlemen was drunk with much warmth of feeling by the artificers and seamen, who severally expressed the satisfaction they had experienced in acting under them; after which the whole party left the rock.

In sailing past the floating light mutual compliments were made by a display of flags between that vessel and the tender; and at five p.m. the latter vessel entered the harbour of Arbroath, where the party were heartily welcomed by a numerous company of spectators, who had collected to see the artificers arrive after so long an absence from the port. In the evening the writer invited the foremen and captains of the service, together with Mr. David Logan, clerk of works at Arbroath, and Mr. Lachlan Kennedy, engineer's clerk and bookkeeper, and some of their friends, to the principal inn, where the evening was spent very happily; and after 'His Majesty's Health' and 'The Commissioners of the Northern Lighthouses' had been given, 'Stability to the Bell Rock Lighthouse' was hailed as a standing toast in the Lighthouse service.

Sunday, 5th Aug.

The author has formerly noticed the uniformly decent and orderly deportment of the artificers who were employed at the Bell Rock Lighthouse, and to-day, it is believed, they very generally attended church, no doubt with grateful hearts for the narrow escapes from personal danger which all of them had more or less experienced during their residence at the rock.

1810 Tuesday, 14th Aug.

The *Smeaton* sailed to-day at one p.m., having on board sixteen artificers, with Mr. Peter Logan, together with a supply of provisions and necessaries, who left the harbour pleased and happy to find themselves once more afloat in the Bell Rock service. At seven o'clock the tender was made fast to her moorings, when the artificers landed on the rock and took possession of their old quarters in the beacon-house, with feelings very different from those of 1807, when the works commenced.

The barometer for some days past had been falling from 29·90, and to-day it was 29·50, with the wind at N.E., which, in the course of this day, increased to a strong gale accompanied with a sea which broke with great violence upon the rock. At twelve noon the tender rode very heavily at her moorings, when her chain broke at about ten fathoms from the ship's bows. The kedge-anchor was immediately let go, to hold her till the floating buoy and broken chain should be got on board. But while this was in operation the hawser of the kedge was chafed through on the rocky bottom and parted, when the vessel was again adrift. Most fortunately, however, she cast off with her head from the rock, and narrowly cleared it, when she sailed up the Firth of Forth to wait the return of better weather. The artificers were thus left upon the rock with so heavy a sea running that it was ascertained to have risen to a height of eighty feet on the building. Under such perilous circumstances it would be difficult to describe the feelings of those who, at this time, were cooped up in the beacon in so forlorn a situation, with the sea not only raging under them, but occasionally falling from a great height upon the roof of their temporary lodging, without even the attending vessel in view to afford the least gleam of hope in the event of any accident. It is true that they now had the masonry of the lighthouse to resort to, which, no doubt, lessened the actual danger of their situation; but the building was still without a roof, and the deadlights, or storm-shutters, not being yet fitted, the windows of the lower story were stove in and broken, and at high-water the sea ran in considerable quantities out at the entrance door.

Thursday, 16th Aug.

The gale continues with unabated violence to-day, and the sprays rise to a still greater height, having been carried over the masonry of the building, or about ninety feet above the level of the sea. At four o'clock this morning it was breaking into the cook's berth, when he rang the alarm-bell, and all hands turned out to attend to their personal safety. The floor of the smith's, or mortar gallery, was now completely burst up by the force of the sea, when the whole of the deals and the remaining articles upon the floor were swept away, such as the cast-iron mortar-tubs, the iron hearth of the forge, the smith's bellows, and even his anvil were thrown down upon the rock. Before the tide rose to its full height to-day some of the artificers passed along the bridge into the lighthouse, to observe the effects of the sea upon it, and they reported that they had felt a slight tremulous motion in the building when great seas struck it in a certain direction, about high-water mark. On this occasion the sprays were again observed to wet the balcony, and even to come over the parapet wall into the interior of the light-room.

Thursday, 23rd Aug.

The wind being at W.S.W., and the weather more moderate, both the tender and the *Smeaton* got to their moorings on the 23rd, when all hands were employed in transporting the sash-frames from on board of the *Smeaton* to the rock. In the act of

1810 setting up one of these frames upon the bridge, it was unguardedly suffered to lose its balance, and in saving it from damage Captain Wilson met with a severe bruise in the groin, on the seat of a gun-shot wound received in the early part of his life. This accident laid him aside for several days.

Monday, 27th Aug. The sash-frames of the light-room, eight in number, and weighing each 254 pounds, having been got safely up to the top of the building, were ranged on the balcony in the order in which they were numbered for their places on the top of the parapet-wall; and the balance-crane, that useful machine having now lifted all the heavier articles, was unscrewed and lowered, to use the landing-master's phrase, 'in mournful silence.'

Sunday, 2nd Sept. The steps of the stair being landed, and all the weightier articles of the light-room got up to the balcony, the wooden bridge was now to be removed, as it had a very powerful effect upon the beacon when a heavy sea struck it, and could not possibly have withstood the storms of a winter. Everything having been cleared from the bridge, and nothing left but the two principal beams with their horizontal braces, James Glen, at high-water, proceeded with a saw to cut through the beams at the end next the beacon, which likewise disengaged their opposite extremity, inserted a few inches into the building. The frame was then gently lowered into the water, and floated off to the *Smeaton* to be towed to Arbroath, to be applied as part of the materials in the erection of the lightkeepers' houses. After the removal of the bridge, the aspect of things at the rock was much altered. The beacon-house and building had both a naked look to those accustomed to their former appearance; a curious optical deception was also

1810

remarked, by which the lighthouse seemed to incline from the perpendicular towards the beacon. The horizontal rope-ladder before noticed was again stretched to preserve the communication, and the artificers were once more obliged to practise the awkward and straddling manner of their passage between them during 1809.

At twelve noon the bell rung for prayers, after which the artificers went to dinner, when the writer passed along the rope-ladder to the lighthouse, and went through the several apartments, which were now cleared of lumber. In the afternoon all hands were summoned to the interior of the house, when he had the satisfaction of laying the upper step of the stair, or last stone of the building. This ceremony concluded with three cheers, the sound of which had a very loud and strange effect within the walls of the lighthouse. At six o'clock Mr. Peter Logan and eleven of the artificers embarked with the writer for Arbroath, leaving Mr. James Glen with the special charge of the beacon and railways, Mr. Robert Selkirk with the building, with a few artificers to fit the temporary windows to render the house habitable.

Sunday, 14th Oct.

On returning from his voyage to the Northern Lighthouses, the writer landed at the Bell Rock on Sunday, the 14th of October, and had the pleasure to find, from the very favourable state of the weather, that the artificers had been enabled to make great progress with the fitting-up of the light-room.

Friday, 19th Oct.

The light-room work had proceeded, as usual, to-day under the direction of Mr. Dove, assisted in the plumber-work by Mr. John Gibson, and in the brazier-work by Mr. Joseph Fraser; while Mr. James Slight, with the joiners, were fitting up the storm-shutters

1810 of the windows. In these several departments the artificers were at work till seven o'clock p.m., and it being then dark, Mr. Dove gave orders to drop work in the light-room; and all hands proceeded from thence to the beacon-house, when Charles Henderson, smith, and Henry Dickson, brazier, left the work together. Being both young men, who had been for several weeks upon the rock, they had become familiar, and even playful, on the most difficult parts about the beacon and building. This evening they were trying to outrun each other in descending from the light-room, when Henderson led the way; but they were in conversation with each other till they came to the rope-ladder distended between the entrance-door of the lighthouse and the beacon. Dickson, on reaching the cook-room, was surprised at not seeing his companion, and inquired hastily for Henderson. Upon which the cook replied, 'Was he before you upon the rope-ladder?' Dickson answered, 'Yes; and I thought I heard something fall.' Upon this the alarm was given, and links were immediately lighted, with which the artificers descended on the legs of the beacon, as near the surface of the water as possible, it being then about full tide, and the sea breaking to a considerable height upon the building, with the wind at S.S.E. But, after watching till low-water, and searching in every direction upon the rock, it appeared that poor Henderson must have unfortunately fallen through the rope-ladder, and been washed into the deep water.

The deceased had passed along this rope-ladder many hundred times, both by day and night, and the operations in which he was employed being nearly finished, he was about to leave the rock when this melancholy catastrophe took place. The

1810

unfortunate loss of Henderson cast a deep gloom upon the minds of all who were at the rock, and it required some management on the part of those who had charge to induce the people to remain patiently at their work; as the weather now became more boisterous, and the nights long, they found their habitation extremely cheerless, while the winds were howling about their ears, and the waves lashing with fury against the beams of their insulated habitation.

Tuesday, 23rd Oct.

The wind had shifted in the night to N.W., and blew a fresh gale, while the sea broke with violence upon the rock. It was found impossible to land, but the writer, from the boat, hailed Mr. Dove, and directed the ball to be immediately fixed. The necessary preparations were accordingly made, while the vessel made short tacks on the southern side of the rock, in comparatively smooth water. At noon Mr. Dove, assisted by Mr. James Slight, Mr. Robert Selkirk, Mr. James Glen, and Mr. John Gibson, plumber, with considerable difficulty, from the boisterous state of the weather, got the gilded ball screwed on, measuring two feet in diameter, and forming the principal ventilator at the upper extremity of the cupola of the light-room. At Mr. Hamilton's desire, a salute of seven guns was fired on this occasion, and, all hands being called to the quarter-deck, 'Stability to the Bell Rock Lighthouse' was not forgotten.

Tuesday, 30th Oct.

On reaching the rock it was found that a very heavy sea still ran upon it; but the writer having been disappointed on two former occasions, and, as the erection of the house might now be considered complete, there being nothing wanted externally, excepting some of the storm-shutters for the defence of the windows, he was the more anxious at this time

1810 to inspect it. Two well-manned boats were therefore ordered to be in attendance; and, after some difficulty, the wind being at N.N.E., they got safely into the western creek, though not without encountering plentiful sprays. It would have been impossible to have attempted a landing to-day, under any other circumstances than with boats perfectly adapted to the purpose, and with seamen who knew every ledge of the rock, and even the length of the sea-weeds at each particular spot, so as to dip their oars into the water accordingly, and thereby prevent them from getting entangled. But what was of no less consequence to the safety of the party, Captain Wilson, who always steered the boat, had a perfect knowledge of the set of the different waves, while the crew never shifted their eyes from observing his motions, and the strictest silence was preserved by every individual except himself.

On entering the house, the writer had the pleasure to find it in a somewhat habitable condition, the lower apartments being closed in with temporary windows, and fitted with proper storm-shutters. The lowest apartment at the head of the staircase was occupied with water, fuel, and provisions, put up in a temporary way until the house could be furnished with proper utensils. The second, or light-room store, was at present much encumbered with various tools and apparatus for the use of the workmen. The kitchen immediately over this had, as yet, been supplied only with a common ship's caboose and plate-iron funnel, while the necessary cooking utensils had been taken from the beacon. The bedroom was for the present used as the joiners' workshop, and the strangers' room, immediately under the light-room, was occupied by the artificers, the beds being ranged in tiers, as was

1810

done in the barrack of the beacon. The light-room, though unprovided with its machinery, being now covered over with the cupola, glazed and painted, had a very complete and cleanly appearance. The balcony was only as yet fitted with a temporary rail, consisting of a few iron stanchions, connected with ropes; and in this state it was necessary to leave it during the winter.

Having gone over the whole of the low-water works on the rock, the beacon, and lighthouse, and being satisfied that only the most untoward accident in the landing of the machinery could prevent the exhibition of the light in the course of the winter, Mr. John Reid, formerly of the floating light, was now put in charge of the lighthouse as principal keeper; Mr. James Slight had charge of the operations of the artificers, while Mr. James Dove and the smiths, having finished the frame of the light-room, left the rock for the present. With these arrangements the writer bade adieu to the works for the season. At eleven a.m. the tide was far advanced; and there being now little or no shelter for the boats at the rock, they had to be pulled through the breach of sea, which came on board in great quantities, and it was with extreme difficulty that they could be kept in the proper direction of the landing-creek. On this occasion he may be permitted to look back with gratitude on the many escapes made in the course of this arduous undertaking, now brought so near to a successful conclusion.

Monday, 5th Nov.

On Monday, the 5th, the yacht again visited the rock, when Mr. Slight and the artificers returned with her to the workyard, where a number of things were still to prepare connected with the temporary fitting up of the accommodation for the lightkeepers. Mr. John Reid and Peter Fortune were now the only

1810 inmates of the house. This was the smallest number of persons hitherto left in the lighthouse. As four lightkeepers were to be the complement, it was intended that three should always be at the rock. Its present inmates, however, could hardly have been better selected for such a situation; Mr. Reid being a person possessed of the strictest notions of duty and habits of regularity from long service on board of a man-of-war, while Mr. Fortune had one of the most happy and contented dispositions imaginable.

Tuesday, 13th Nov.

From Saturday the 10th till Tuesday the 13th, the wind had been from N.E. blowing a heavy gale; but to-day, the weather having greatly moderated, Captain Taylor, who now commanded the *Smeaton*, sailed at two o'clock a.m. for the Bell Rock. At five the floating light was hailed and found to be all well. Being a fine moonlight morning, the seamen were changed from the one ship to the other. At eight, the *Smeaton* being off the rock, the boats were manned, and taking a supply of water, fuel, and other necessaries, landed at the western side, when Mr. Reid and Mr. Fortune were found in good health and spirits.

Mr. Reid stated that during the late gales, particularly on Friday, the 30th, the wind veering from S.E. to N.E., both he and Mr. Fortune sensibly felt the house tremble when particular seas struck, about the time of high-water; the former observing that it was a tremor of that sort which rather tended to convince him that everything about the building was sound, and reminded him of the effect produced when a good log of timber is struck sharply with a mallet; but, with every confidence in the stability of the building, he nevertheless confessed that, in so forlorn a situation, they were not insensible to those emotions which, he emphatically observed, 'made a man look back upon his former life.' 1810

The day, long wished for, on which the mariner was to see a light exhibited on the Bell Rock at length arrived. Captain Wilson, as usual, hoisted the float's lanterns to the topmast on the evening of the 1st of February; but the moment that the light appeared on the rock, the crew, giving three cheers, lowered them, and finally extinguished the lights. 1811 Friday, 1st Feb.

Printed by T. and A. Constable, Printers to His Majesty
at the Edinburgh University Press

Printed in the United States
By Bookmasters